Micro and Nano Machined Electrometers

Yong Zhu
Editor

Micro and Nano Machined Electrometers

Springer

Editor
Yong Zhu
School of Engineering and Built
Environment
Griffith University, Gold Coast Campus
Southport, QLD, Australia

ISBN 978-981-13-3246-3 ISBN 978-981-13-3247-0 (eBook)
https://doi.org/10.1007/978-981-13-3247-0

This Springer imprint is published by the registered company Springer Nature Singapore Pte Ltd.
The registered company address is: 152 Beach Road, #21-01/04 Gateway East, Singapore 189721, Singapore

Contents

Micromachined Resonant Electrometers

Jin Xie and Dongyang Chen

Abstract This chapter introduces ultrasensitive charge measurement by frequency modulation on micromachined resonators and oscillators at room temperature. The frequency modulation type resonant electrometers have merits of high resolution and large dynamic range. Fundamentals of vibration system is firstly introduced, including resonance of a beam resonator. Oscillations of the resonator with closed loop methods are introduced to obtain and track resonant frequency. Designs, principles and models of several capacitive resonant electrometers are presented. The axial strain modulation and lateral stiffness perturbation charge measurement schemes are studied respectively. Finally, we introduce a prototype of resonant electrometer employing single anchored circular beam resonator with parallel-plate capacitors as transducers based on lateral stiffness perturbation scheme. The prototype of resonant electrometer has sub-electron resolution and high charge sensitivity. The micromachined resonant electrometers may be key components to develop advanced instruments, such as voltmeters, ammeters and multimeters.

1 Introduction

Charge is a quantifiable physical quantity. The least quantized charge called elementary charge with unit value of 1.602×10^{-19} C, usually carried by individual proton or electron. Substances in our natural world exhibit electro-positive or negative characteristics as a physical property in terms of the unbalanced numbers of electrons and protons. Electron is one of the elementary particle to constitute natural matters, so measurement of electric charge is vital. Besides, most of the electrical quantities are other existing forms of charge, e.g. the electrical current is produced by charge flow.

An electrometer is a type of sensor that can precisely measure small or large quantities of electric charge. Thus, it plays an important role in scientific researches

J. Xie (✉) · D. Chen
State Key Laboratory of Fluid Power and Mechatronic Systems, Zhejiang University, Hangzhou, Zhejiang 310027, People's Republic of China
e-mail: xiejin@zju.edu.cn

Y. Zhu (ed.), *Micro and Nano Machined Electrometers*,
https://doi.org/10.1007/978-981-13-3247-0_1

and industrial applications, such as biological and chemical charge detections [1, 2], mass spectrometry [3], nuclear physics [4], high-end instruments e.g. scanning tunneling microscope [5]. To date, high performance electrometry has been explored for a long history and a variety of macro and micro electrometers with different kinds of charge measurement scheme have been developed [5–29], dating back from Coulomb's torsion-balance electrometer more than 200 years ago to the popular gold leaf electroscope and Kelvin's quadrant electrometer, and to nowadays transistors, carbon nanotubes and quantum dot devices, etc. Currently, the mesoscopic single electron transistors can offer ultrahigh resolution below 10^{-5} e/$\sqrt{}$Hz, but the devices also suffer from limited bandwidth meanwhile are operative only at cryogenic temperatures because the underlying principle of Coulomb blockade is prone to be suppressed by thermal fluctuations [8, 9]. Nanoscale mechanical electrometers can also perform ultra-sensitive electrometry approaching quantum limits, but still, requires an operating temperature below 4.2 K [17, 18]. Overall, electrometry with electron level at room temperature is quite limited.

To overcome the above problems, a concept of resonant electrometry for measuring charge at electron level under room temperature has been proposed base on micro and nano electromechanical system (M/NEMS) technologies. Basically, the concept of resonant electrometry refers to a metrology that measures charge with the utilization of resonators and oscillators as the core sensitive element and their resonant frequency as the output metric of interest. A resonator is a type of device that can oscillate with greater amplitude at resonant frequency than other frequencies. An oscillator includes external energy replenishment to a resonator to compensate the energy loss per periodic cycle thus keep a stable periodic motion sustained. Based on that, resonant electrometers can convert the perturbation of input charge to the change in output periodical signal of the core resonator oscillators, e.g. frequency. With miniaturization by micro and nano machining techniques, the size and power consumption of the electrometers are reduced.

Three types of the electrometers employing micromachined resonators based on different charge sensitive principles and output metrics have been presented by researchers. The vibrating reed electrometers [19–22] utilize the variable capacitances generated by resonant motion of a resonator to modulate input charge to an AC voltage output at twice of the operating frequency of the resonator. The output at higher frequencies can reduce the influences from low frequency noises, such as $1/f$ noise. Mode localized electrometers [23–25] output amplitude ratio as metric based on the principle of vibration energy localization between weakly coupled mechanical resonators. Charge sensitivity is boosted by hundreds of times comparing to that of each resonator in the coupled array which monitors resonant frequency as output metric. Resonance or oscillation states are required in vibrating reed and mode localized paradigms during measuring charge, and the amplitudes are more concerned rather than other output features. Detailed introductions on the principles and performances of these two charge measurement paradigms can be found in other chapters of this book. This chapter focuses on the type of resonant sensing which is based on frequency modulation of the resonators and oscillators [25–29].

A variety of techniques for expeditiously and precisely tracking high frequencies have been highly developed, so that the use of the conversion between input charge and resonant frequency as measurement scheme is of particular interest. Especially when it comes to mechanical vibration system, the resonance from silicon resonators with high quality factors also offer ultrahigh stability approaching ppm to ppb level of precision [30] while the same level of precision in voltage or current monitoring is a challenge. The resonant electrometers have merits of high resolution and large dynamic range. This type of sensing paradigm also has advantages of simple structures and implementations, quasi-nature of frequency output for simplifying post-processing circuitry, e.g. the quasi-digital signal does not need an analog to digital converter, which is widely included in the sensor circuitry that monitors amplitudes or eigenstates.

This chapter introduces ultrasensitive charge measurement approaching single electron resolution by frequency modulation on micromachined resonators and oscillators at room temperature. The detailed mechanism of electrometry by frequency modulation is elaborated with background knowledge of resonance and oscillation in the mechanical vibration system and the electrical oscillatory loop. Several typical resonant electrometers are demonstrated with different frequency modulation methods as charge measurement schemes.

2 Vibration System

As a key component in the resonant electrometry, resonators function as an energy storing element while being embedded in a constructed vibration system and is sensitive to external energy injection pertaining to charge input. In such context, mechanical resonators are preferred due to the extraordinary quality factors they offer [31–33], usually exceeding electronic counterparts over several orders of magnitude. With high quality factor, the input energy is preserved in a narrow bandwidth, and thus a good signal to noise ratio and frequency stability can be achieved. For mechanical resonators, two operation states should be clearly understood, the resonance and the oscillation. Resonance is the state that the resonator is operated at the natural frequency and usually accompanied with maximum vibrating amplitude among other frequencies. In addition, each natural frequency has specified mode shape pertaining to particular pattern of motion of the mechanical complaint structures. Oscillation refers to the autonomous motion state generally constructed from resonators embedded in a self-sustained loop. In terms of resonant sensing, those states are essential and will be introduced in this section.

2.1 Resonance

Flexural mechanical beams are the vibratory elements that are frequently employed in resonant electrometers. In the resonant motions, either in the forms of in-plane flexural mode [34] or out-of-plane torsional mode vibrations [35], the flexural mechanical beams can be described as a mass-spring-damper vibration system, as shown in Fig. 1. The dynamic response of a single-degree-of-freedom linear second order mechanical resonator can be formulated by an ordinary differential equation

$$m\ddot{u}(t) + c\dot{u}(t) + ku(t) = F(t) \tag{1}$$

where $u(t)$ represents the time dependent displacement of the mass m, the vibratory element in the resonator, e.g. the straight beams. The motion of the beam at resonance is assumed to be harmonic given by $u(t) = X_A \sin(\omega_n t - \varphi)$ in which t is the time and X_A is the vibration amplitude at the angular natural frequency ω_n; φ is the phase determined by initial state of the resonator. The dot denotes the time derivative. k is the mechanical stiffness of the resonator. c is the damping coefficient. $F(t)$ is the time dependent dynamic force to drive the resonator in the vicinity of its natural frequency ω_n. To solve the equation, the driving force $F(t)$ is assumed to be balanced by the damping of the vibration system, in that case the resonance can be sustained. Therefore, Eq. (1) turns to a homogeneous differential form, given by

$$m\ddot{u}(t) + ku(t) = 0 \tag{2}$$

Then the harmonic displacement $u(t)$ is substituted into Eq. (2), and the analytical solution of the natural angular frequency ω_n is obtained

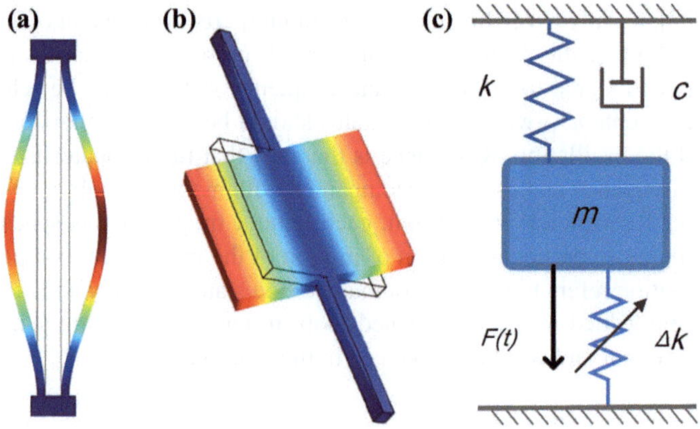

Fig. 1 **a** In plane flexural vibration mode of a beams resonator. **b** Out of plane torsional vibration mode of a beam resonator. **c** Mass-spring-damper system to describe the resonant motion of the mechanical resonators

$$\omega_n = \sqrt{\frac{k}{m}} \tag{3}$$

Equation (3) indicates that the frequency response of a resonator can be modulated by changing mass or stiffness. For resonant sensing, the quantities to be detected can always be transformed from the change of these two parameters. For instances, the former one is mainly used in the measurements of small amount of mass, e.g. molecules or particulate matters [36, 37], etc. The latter one is frequently used in the sensor applications based on force detection, e.g. inertial forces by acceleration or angular rate [38], electrostatic forces by electric charge or voltage [26–28] and magnetic forces in magnetic field [39]. Commonly, in resonant electrometry, the electrostatic force generated from input charge produces a stiffness change of $\pm\Delta k$ to the initial mechanical stiffness k of the resonator, as described by Fig. 1c, which causes strengthening or weakening of the overall stiffness and thus changes the natural frequency.

Firstly, the resonant motion without stiffness modulation is considered. Depending on the choices of transducers [40–42], resonators can be driven into resonance electrostatically, piezoelectrically or thermally in the vicinity of the natural frequency. Assume the forcing term is sinusoidal given by $F(t) = F\sin(\Omega t)$, where Ω is the excitation frequency of the driving force F, then the displacement response $u(t)$ will also be sinusoidal. Equation (1) can be rewritten in the form as follows

$$\ddot{u}(t) + 2\zeta\omega_n\dot{u}(t) + \omega_n^2 u(t) = \frac{F\sin(\Omega t)}{m} \tag{4}$$

where ζ is the damping ratio of the vibration system, defined as $\zeta = c/2m\omega_n$, which can also be expressed by $\zeta = 1/2Q$. Q is the quality factor. Equation (4) can be solved by using complex functions to represent the sinusoidal terms, e.g. the force and displacement terms. As a result, the amplitude and phase responses of the vibration system regarding to the frequency of the driving force Ω can be derived as

$$X_A(\Omega) = \frac{F/k}{\sqrt{(1-(\Omega/\omega_n)^2)^2 + 4\zeta^2(\Omega/\omega_n)^2}}, \quad \varphi(\Omega) = \arctan\left(\frac{2\zeta\Omega/\omega_n}{1-(\Omega/\omega_n)^2}\right) \tag{5}$$

According to Eq. (5), the amplitude and phase functions are plotted in Fig. 2a, b regarding to the normalized forcing frequency Ω/ω_n. While the excitation frequency deviates from the natural frequency of the resonator, the amplitude response approaches static displacement $X_A(0) = F/k$, wherein the resonator simply acts as a static spring. But in the vicinity of natural frequency, the amplitude responses are far greater than static displacement, especially in the condition of relative lower damping ratio. As a result of the damping, the peak amplitude of the resonance does not occur at the natural frequency but near it, given by

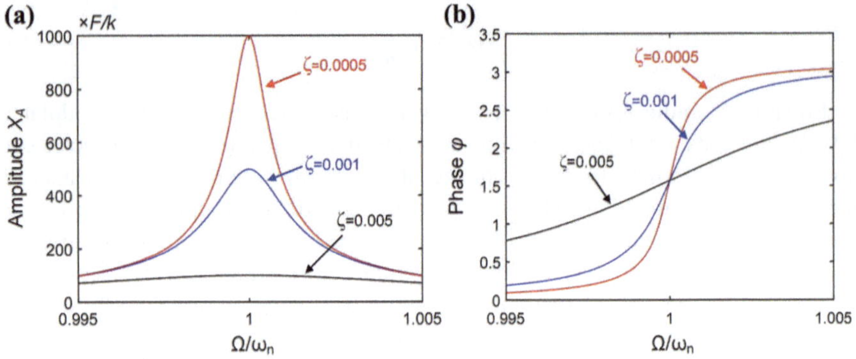

Fig. 2 **a** The amplitude and **b** phase functions of the single degree of freedom linear second order resonators under different damping ratios. Lower damping ratio indicates higher quality factor; thus, the frequency spectrum has narrower bandwidth and higher peak amplitude

$$X_A\left(\omega_n\sqrt{1-\zeta^2}\right)=\frac{F/k}{\sqrt{4\zeta^2\sqrt{1-\zeta^2}}} \tag{6}$$

Accordingly, the angular resonant frequency ω_r is determined by $\omega_r = \omega_n\sqrt{1-\zeta^2}$. Generally, a higher peak amplitude accompanied with narrower bandwidth will arise in the spectra by decreasing the damping ratio or increasing the quality factor. A lightly damped resonator has a mechanical resonant frequency very close to the natural frequency. Taking account of stiffness modulation during charge measurement, the modulated mechanical stiffness of the resonator yields a change in frequency responses, as can be modelled by

$$\omega_r=\sqrt{\frac{k+\Delta k}{m}}\sqrt{1-\zeta^2}=\sqrt{1+\frac{\Delta k}{k}}\sqrt{1-\zeta^2}\omega_n \tag{7}$$

Equation (7) indicates that the stiffness modulation will lead to upward or downward resonant frequency shift depending on the sign of Δk. Combining Eqs. (5) and (7), the periodic solution of the resonant motion can be found, as shown in Fig. 3. Subject to a stiffness modulation of $\pm 1‰$, the opposite translations of the frequency spectra are observed, including both amplitude and phase responses. This shows obvious opposite shifts of resonant frequency depending on positive or negative stiffness induced by the charge being measured. Although the resonant frequency at peak amplitude changes under stiffness modulation, the phase at that point remains unchanged. In other words, the relation between frequency, amplitude and phase does not change with the stiffness modulation. This indicates a determined correspondence in frequency, amplitude and phase during the charge measurement by resonant sensing. By locking the phase or controlling the amplitude of this periodic signal in an oscillation loop, the frequency has unique solution so that the real

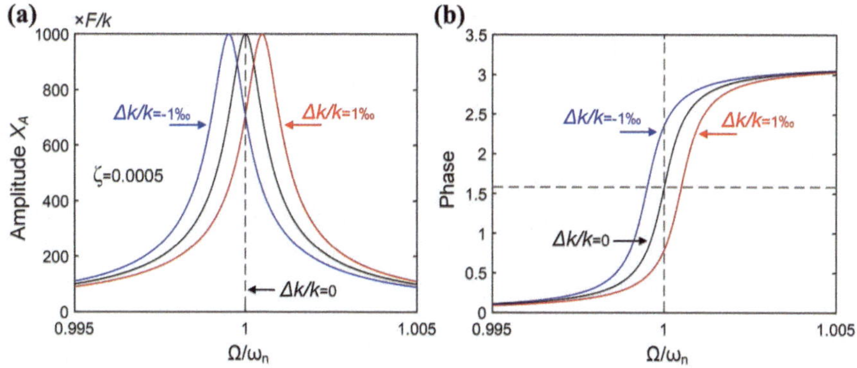

Fig. 3 The opposite changes of **a** amplitude and **b** phase responses in frequency spectra in response to the positive and negative stiffness modulation to the resonant vibration system

time tracking on the frequency response in electrometry is achievable, which will be introduced in Sect. 2.2.

It should be pointed out that the stiffness modulation equally gives rise to the frequency change throughout the resonance bandwidth. The peak amplitude at resonant frequency is chosen as output because of the best signal and power handling level, and the lowest phase noise to frequency noise conversion. Periodical signal in real world always accompanies with many kinds of noises. As for a sinusoidal signal transduced from mesoscopic periodical motion at a fixed frequency, as shown in Fig. 4, the phase noise dominates frequency fluctuation of the linear resonators while amplitude noise does not influence on the frequency. The derivative of the phase function with respect to frequency is depicted in Fig. 5, where the derivative $\partial\varphi/\partial(\Omega/\omega_n)$ reaches maximum at peak amplitude at resonant frequency ω_r.

Fig. 4 Sinusoidal signal responses of the resonator at a fixed Ω/ω_n in time domain with amplitude and phase noises

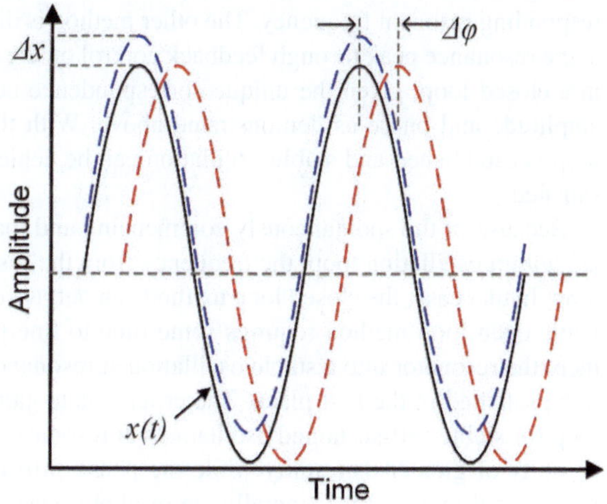

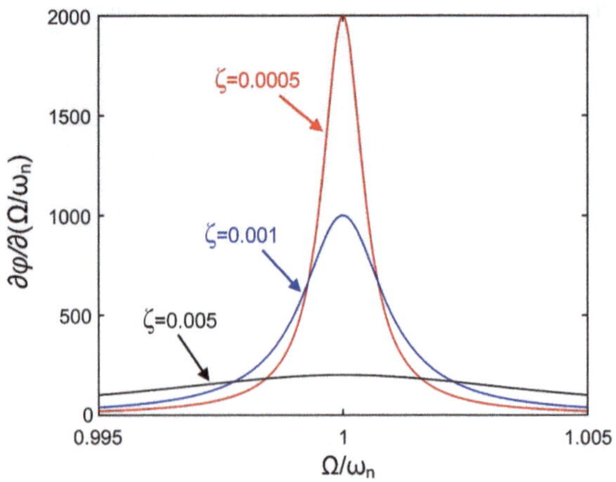

Fig. 5 Ω/ω_n derivation of the phase function in Fig. 2b under different damping ratios to show the phase-frequency conversion in resonant motion

Therefore, another advantage of choosing resonant frequency ω_r as output of the measurements is the minimum frequency noise conversion from phase fluctuation.

2.2 *Oscillation*

For the resonant electrometry by frequency modulation, the resonant frequency ω_r is of particular interest since it can accurately measure charge quantities with less noise interferences and better signal to noise ratio. In an attempt to obtain the resonant frequency of the resonator, two methods are commonly used. The first one is mapping the frequency spectra by using continuous Ω to sweep the resonator in the vicinity of the natural frequency to get the amplitude responses in an open loop implementation. During the process, the maximum amplitude can be then extracted as well as the corresponding resonant frequency. The other method is directly operating the resonator at the resonance peak through feedback control on the resonance amplitude or phase in a closed loop, given the unique correspondence between resonant frequency to amplitude and phase as demonstrated above. With the feedback control, a closed loop is established and stable oscillation can be achieved while several criteria are satisfied.

Because of the spontaneously commencing and maintaining of the periodic signal within oscillation loop, the frequency from the resonant motion is locked in the loop. In that case, the closed loop method can output resonant frequency in real time while open loop method requires some time to finely sweep frequency. To implement the resonator into a stable oscillation at resonance, Barkhausen's criteria need to be satisfied in the first place. The criteria state gain and phase conditions of the loop for stable self-sustained oscillations at resonance. The gain of the loop should be unity or greater than unity while the phase shift around the loop should be an integer multiple of 2π. Generally, an oscillator system with least components can break down to a resonator incorporating with a sustaining amplifier. The sustaining amplifier can compensate the energy loss of the mechanical resonator at resonance so that the loop gain condition is satisfied. But most of the time the sustaining amplifier will introduce unwanted phase error to the loop as well, so additional components are required to satisfy the phase condition, e.g. a phase shifter. Barkhausen's criteria are necessary conditions for oscillation, when it comes to the practical implementation of oscillators with capacitive, piezoelectric or thermal transducers for resonant motion actuation and monitoring, more criteria will be required. Capacitively driven resonators are taken as examples, which are frequently used in resonant electrometry and other sensing applications. The electrical model for the resonant system should be formulated for placing insights in building a resonator oscillator.

The description of the resonant motion in mechanical domain can be equivalently translated into electrical domain according to the relations: the mechanical concepts of damping c, mass m and compliance $1/k$ are equivalent to the electrical components of resistance R, inductance L and capacitance C, respectively. The link between mechanical and electrical models of the resonator is electromechanical coupling factor η of the capacitive transducer. η is defined as the ratio of the displaced charge $(\partial C_P \times V_P)$ over the spatial displacement of the resonant motion. Therefore, the translations of these parameters are given by

$$\eta = \frac{\partial C_P}{\partial x} V_P, \; R = \frac{c}{\eta^2}, \; L = \frac{m}{\eta^2}, \; C = \frac{\eta^2}{k} \tag{8}$$

where C_P denotes the capacitance of the transducer. V_P is the polarization voltage imposed on the transducer for providing charge coupling. η determines the conversion of the stiffness in mechanical system to the series capacitance in RLC loop. Therefore, the stiffness modulation Δk can be translated into capacitance variation ΔC. However, in capacitive mechanical resonators, the use of parallel-plate or comb capacitors as transducers will also create certain feedthrough passage. In that case, the mechanical resonance with feedthrough effect can be described by incorporating series RLC with a parallel feedthrough capacitor C_f, as depicted in Fig. 6. The feedthrough capacitor represents the capacitive paths through which the current from driving signal directly feeds to the output ports. Note that feedthrough effect is a common phenomenon existing in miniaturized capacitive, piezoelectric and thermal transducers. To analyze the feedthrough effect in the loop, the admittance of the resonator, which is proportional to the resonant amplitude, is modelled as

$$Y(\omega) = \frac{1}{R + j\omega L + 1/(j\omega C)} \pm j\omega C_f \tag{9}$$

where ω is the excitation frequency equivalent to Ω in the mechanical domain. The first and second term in the right-hand side of Eq. (9) are the admittance from series RLC and parallel feedthrough, respectively. The choice of using plus or minus sign in second term is determined by whether the phase of feedthrough current i_f is behind or ahead of the motional current i_s, which can be predicted by the initial design and the

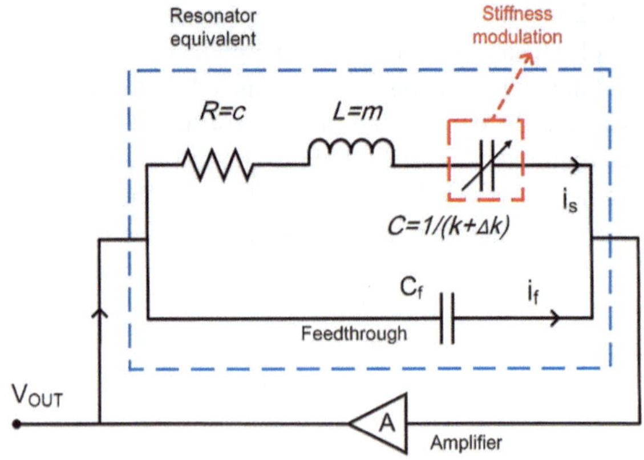

Fig. 6 The equivalent RLC loop for describing the mechanical resonator in oscillation loop. The basic oscillation loop consists of a resonator and a sustaining amplifier

specific polarization manners on the transducers. The natural frequency in electrical domain in the absence of feedthrough effect ($C_f = 0$) is deduced as $\omega_e = 1/\sqrt{LC}$ in terms of that the admittance reaches maximum value of $Y(\omega_e) = 1/R$. ω_e can also be derived by combining Eqs. (3) and (8). Taking account of the feedthrough effect, the resonant frequency will absolutely be influenced by C_f.

Figure 7 depicts the overall admittance of the equivalent RLC loop normalized to $Y(\omega_e) = 1/R$ in the Nyquist plot for different values of ωC_f. The real and imaginary parts of the overall admittance can be drawn as a circle with diameter of $1/R$. The admittance circle is formed by upward sweeping ω in the vicinity of ω_e along the clockwise direction from the origin of the coordinate system. From the origin of the complex plane to each point of the admittance circle, a vector can be drawn e.g. with the solid arrow-line in Fig. 7. The length of the solid arrow-line is the modulus of overall admittance while the angle between the solid arrow-line to horizontal axis represents the phase offset. ω_e no longer has maximum admittance while feedthrough comes into play and shifts the admittance circle upward and downward depending on phase leading or lagging comparing to the resonant motion. As the translation of admittance circle, the modulus of the vector increases, indicating the increase of overall output current from the loop. While ωC_f is far greater than $1/R$, the admittance circle will shift away from original one. By that time the modulus of the vector is dominated by C_f and the phase change is limited in a small range. This situation causes important features of the resonant motion interfered by serious feedthrough component and should be avoided.

The feedthrough absolutely causes resonant frequency of the loop deviating from ω_e and attributes to the change in amplitude and phase. For quantitative analysis, the quality factor of the RLC loop is converted by $Q = 1/(2\zeta) = \sqrt{km}/c = \sqrt{L}/\left(R\sqrt{C}\right)$. Substituting that to Eq. (9), the modulus and phase of the vector is deduced as

$$|Y(\omega)| = \sqrt{\frac{1}{R^2}\frac{1 - 2R\omega C_f Q(\omega/\omega_e - \omega_e/\omega)}{1 + Q^2(\omega/\omega_e - \omega_e/\omega)^2} \pm (\omega C_f)^2}$$

$$\arg(Y) = \arctan(-QR(\omega/\omega_e - \omega_e/\omega)) + \omega C_f\left(R + RQ^2(\omega/\omega_e - \omega_e/\omega)^2\right) \quad (10)$$

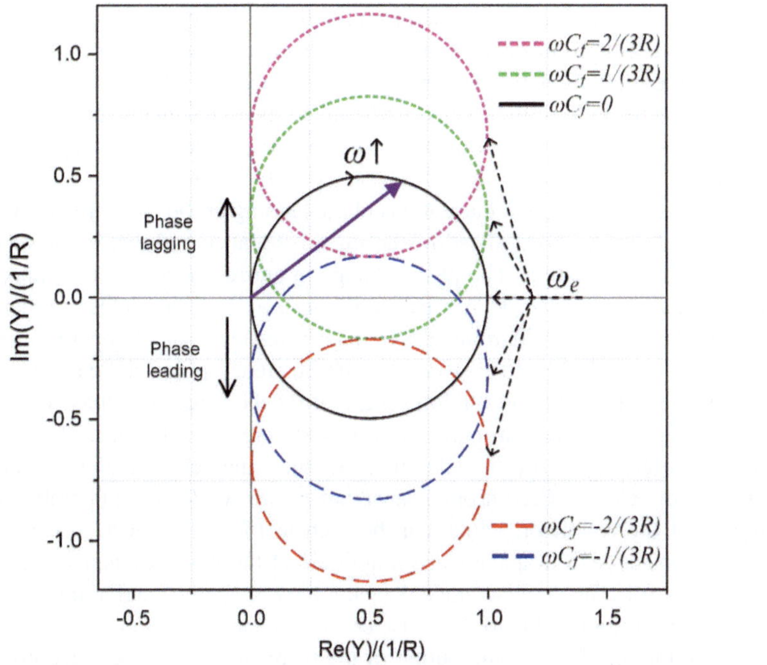

Fig. 7 The overall admittance of the resonator equivalent loop normalized to $1/R$ in the complex plane under different feedthrough admittances ωC_f with phase lagging and leading conditions

According to Eq. (10), the amplitude responses can be obtained, as shown in Fig. 8. The feedthrough current creates antiresonance peak, also known as the parallel resonance, near the series resonance peak of RLC loop. The magnitude of resonant signal at the peak amplitude actually decreases after the feedthrough comes into

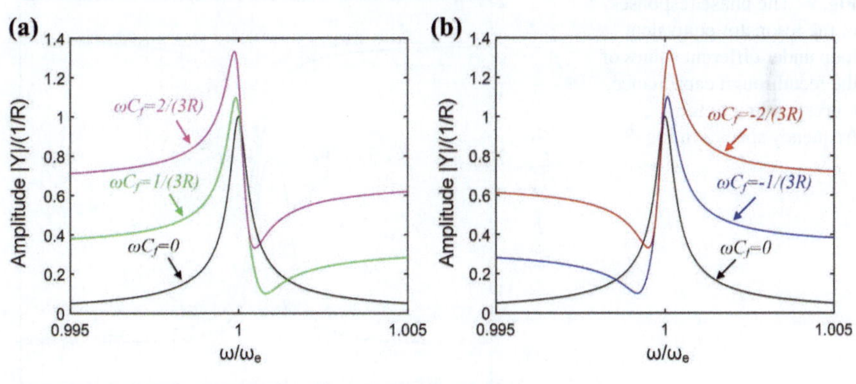

Fig. 8 The amplitude responses of the resonator equivalent loop normalized to $1/R$ with anti-resonance peaks behind (**a**) or ahead (**b**) of series resonance of RLC in terms of different values of the feedthrough capacitance

play. This indicates the motional current at the peak amplitude is partly submerged by the feedthrough current and will be entirely covered while C_f is higher enough. To commence oscillation, the feedthrough capacitance should be as small as possible in the resonators and oscillators in order to avoid the real signal to be covered. Several techniques have been demonstrated in order to reduce feedthrough in capacitive resonators [43–45].

While the feedthrough current has phase leading ahead of the resonance current, the antiresonance arises ahead of the main resonance and that is converse for the phase lagging situation. The phase change in antiresonance is also opposite to the one in main resonance, as can be seen from Fig. 9. That yields another criterion for building stable oscillations: the frequency response should decrease as the loop phase increases. In Fig. 9, each state on the frequency-phase curve under $C_f = 0$ can be operated in a stable oscillation state. But for the frequency-phase curves subject to the influence of feedthrough effect, only the branches that meet the criterion can build oscillations, e.g. the branches that have the same trend with the frequency-phase curve under $C_f = 0$. Feedthrough capacitance narrows down the operable oscillation region of the resonator, which can be seen in Fig. 21 in Sect. 3.3, where stable oscillations at each frequency and amplitude of the main resonance are created by phase control. The antiresonance does not have frequency and amplitude solutions for stable oscillation due to the phase criterion is not met.

Even though the periodic output of the resonators comprises feedthrough component, the change of resonant frequency with respect to charge measurement is not affected, as indicated by Eq. (10). Figure 10 plots frequency spectra of the electrical model, including amplitude and phase responses with antiresonance, while $\pm 5‰$ perturbations on the series capacitance C are introduced to simulate charge measurement. Opposite translations of frequency spectrum can be observed in response to the perturbation. The phase at the resonance peak also remains unchanged despite the change of resonant frequency. In other words, the resonant frequency always

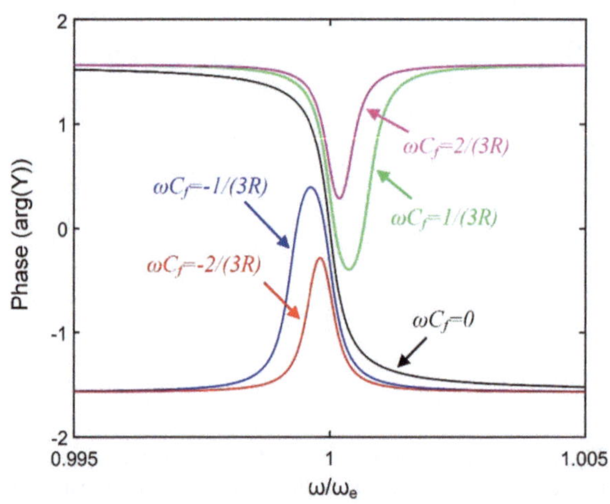

Fig. 9 The phase responses of the resonator equivalent loop under different values of the feedthrough capacitance, corresponding to the frequency spectra in Fig. 8

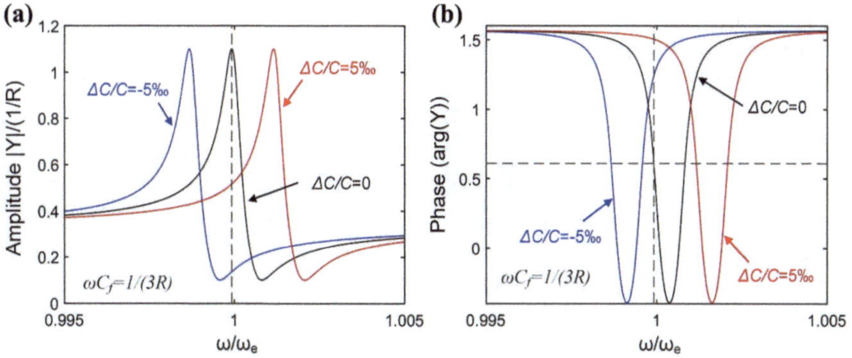

Fig. 10 The opposite changes of **a** amplitude and **b** phase responses in frequency spectra in response to the positive and negative capacitance modulation to the RLC loop with feedthrough effect

appears at maximum amplitude with the same phase offset during measurement. These inherent spectral characteristics pave the foundation for accurately tracking resonant frequency in oscillation state by amplitude or phase control, the same as the conclusion from the above analysis on the mass-spring-damper vibration system. While the criteria for oscillation are satisfied, the resonant frequency of interest can be found and tracked in real time by locking amplitude or phase in an established closed loop. Two types of closed loop configurations are frequently used in resonant electrometers and other resonant sensors, namely the automatic gain control (AGC) and phase locked loop (PLL).

AGC is implemented by self-limiting of vibration amplitude in resonant motion. With the AGC, the resonators work in a linear regime with small amplitude and will not suffer from the nonlinearities [46–48]. In that case, nonlinearities introduced amplitude to frequency noise conversion will not affect oscillator stability. The AGC closes the resonant loop by feeding the output of the resonator to its input port after amplified by electronic amplifiers. The oscillation can be triggered with a white noise input to the loop, then the oscillation amplitude keeps increasing until the loop gain equals to unity, and finally a stable oscillation is maintained with steady amplitude. In this case, the gain of the loop is defined as the multiple of the transfer function of the resonator and the sustaining amplifiers. Even when the loop gain is larger than unity, by employing an amplitude limiting components, e.g. comparators or operational amplifiers, the oscillation amplitude can be kept dropping until the loop gain is lowered to unity [49]. The gain of the amplifier should be a function of the oscillation amplitude, so the loop gain is kept at constant of unity. While the gain of the amplifier decreases, the oscillation amplitude should increase. Following the above summaries for amplitude control of the resonators, many oscillation circuit models have been demonstrated, such as Pierce structure [50] and auto amplitude control [51], etc.

A simple example for employing amplitude controlling oscillation in resonators is depicted in Fig. 11. The oscillation enabling loop employs a transimpedance amplifier

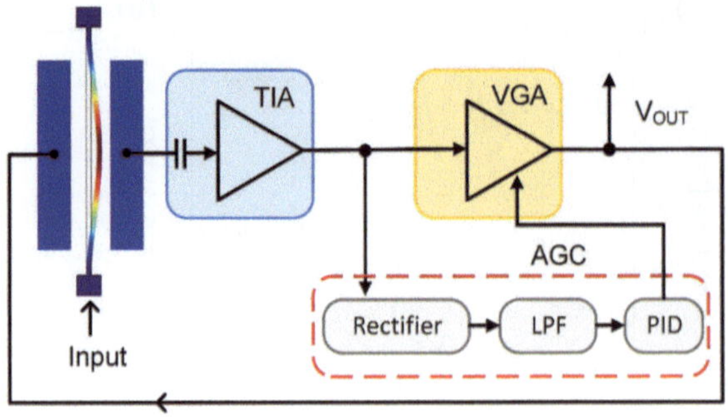

Fig. 11 Osillation loop of resonant sensor employing automatic gain control

(TIA) for converting and amplifying output motional current of the resonator to voltage, a variable gain amplifier (VGA) for keeping a stable gain in the loop, and an AGC loop to get amplitude information for the VGA to control the gain. Generally, the VGA is a voltage-controlled amplifier. To obtain a feedback of DC voltage to VGA, the AGC loop needs to conduct rectification, low-pass filtering (LPF) and proportional-integral-derivative (PID) control on the periodic signal from the TIA. The VGA and AGC together can realize amplitude detection and overall gain control of the loop. As a result, V_{OUT} is a stable periodic signal produced by the oscillation loop and is outputted in real time.

For phase control enabled oscillation, a PLL can track the phase of the output motional current of the resonator and then drive the resonator with certain phase offset by using an external oscillator. Under the phase offset condition between the two ports of the resonator, the frequency of the loop has unique solution according to its inherent spectral characteristics as illustrated above, and thus can be tracked. With the PLL in sensor implementation, the output amplitude can be enhanced by increasing driving amplitude, even beyond the boundary of nonlinearities. In fact, recent researches have demonstrated that operating the oscillators at specific bifurcation points in nonlinear regime can improve phase noise, signal to noise ratio, thus frequency fluctuation and ultimately the resolution of the sensors [52–54].

A simple implementation of oscillation loop with a PLL is depicted in Fig. 12. Different from the AGC loop, the PLL requires a phase detector (PD) for determining the phase of the reference signal outputted from the TIA. Then a voltage-controlled oscillator (VCO) is used to produce a periodic signal with consistent phase by using PID control. Usually, the amplitude value of the periodic signal from VCO can also be adjusted to drive the oscillator toward different vibration amplitudes. A phase shifter is used to control the phase difference between the driving and monitoring ports of the resonator, aiming for creating phase condition in the loop for operating the oscillator in resonance state or other states, e.g. bifurcation in nonlinear regime. The stable oscillation signal that incorporates original output voltage from the resonator is demodulated as V_{OUT} by using the standard signal generated from VCO. V_{OUT} is the periodic signal which has exact resonant amplitude, phase and frequency features of the oscillation loop. When the measurand is the input charge to the oscillator, the oscillation frequency will change to response even though the phase of

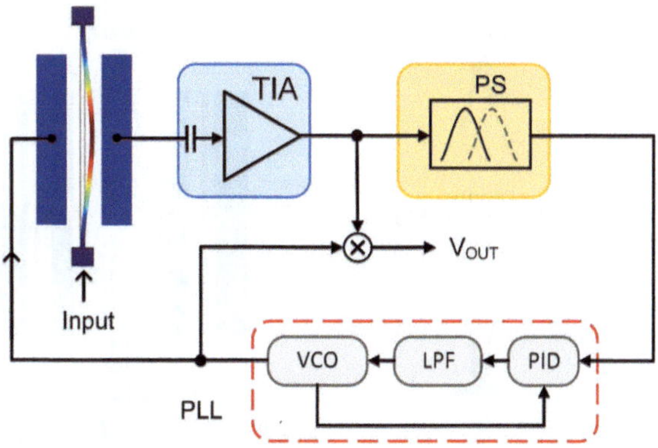

Fig. 12 Oscillation loop of resonant sensor employing phase locked loop

the loop is locked, as explained above. The spectral characteristics of the resonant sensing enables the real time tracking of the oscillation frequency subject to charge measurement.

3 Charge Measurement Paradigms

In resonant electrometry, the physical process of charge measurement can be considered as energy addition and preservation to the resonator vibration system through electrostatic force from the input charge. As afore analyzed, the continuous mutual conversion between kinetic and potential energies occurs during resonance. In terms of potential energy, there are two ways for injecting potential energy: (1) to create elastic change in mechanical vibratory element of the resonator thus alter the potential energy; (2) to influence the resonant motion of the resonator thus perturb the potential energy. Those energy changes in the vibration system can all be concluded as stiffness modulation in mass-spring-damper system or capacitance modulation in RLC loop. Ultimately, the resonant frequency is modulated, which is theoretically analyzed in resonance and oscillation sections.

A typical design of the double ended tuning forks (DETF) resonator is shown in Fig. 13a. The potential energy of the beams can be changed by axially modulating the beams into tensile or compressive states. This charge measurement scheme is called axial strain modulation [24]. The other way to change potential energy is to use electrostatic force to laterally perturb the resonant motion. As a result, an effective electrical stiffness will be added and superimposed on the mechanical stiffness. This method is called lateral stiffness perturbation. Basically, the resonant sensing in electrometry can be concluded as the detection of strain or internal stress in a resonator. Apart from key mechanical resonators, the resonant electrometer design involves the coupling mechanism that can convert the charge into stiffness disturbance. A simple and widely used method is designing a small capacitor aside the vibratory element of the resonator, and the input charge induces an electrostatic force across the capacitor.

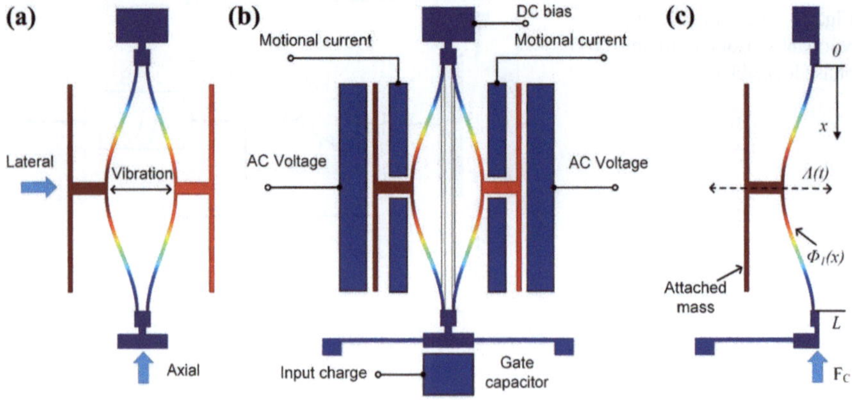

Fig. 13 **a** Schematic of potential energy input paths from axial and lateral directions of a double ended tuning forks resonator. **b** A simple design of resonant electrometer based on axial strain modulation scheme with parallel-plate capacitors for actuation and charge measurement. **c** Vibration analysis on the beams resonator incorporating with generalized modal coordinate and mode shape function

The electrostatic force is exerted on the vibratory element, which changes the elastic property or perturbing the motion state.

This section will elaborate basic designs, principles and models of several resonant electrometers that are based on frequency modulation but with different measurement paradigms. The capacitive transduction mechanism is widely used in those devices in terms of the simple implementation that only DC and AC power sources are needed for motion actuation and detection. That requires electrical conductivity and structure flexibility of the device layer for high performance in electromechanical transduction. The material of the device is mostly silicon because of its mechanical and electrical properties, including the excellent mechanical and chemical stability and the high conductivity after doping. The resonant electrometers being demonstrated here are based on a microelectromechanical system (MEMS) technology and a commercialized fabrication process on silicon-on-insulator (SOI) wafers [55]. Highly doped single crystal silicon is used as device layer and is patterned through deep reaction ion etch (DRIE) process. The vibratory structures in device layer are released and suspended by surface micromachining process. These fabrication techniques are well compatible to nowadays semiconductor and integrating circuit manufacturing [56], so the resonant electrometers have merits of small size, low cost and low power consumption.

3.1 Axial Strain Modulation

The axial strain modulation scheme is frequently implemented by flexural beams resonators. The mechanical beams usually have large length to width ratio to provide an apparent axial direction for the uniformly applied force. A simple design of the resonant electrometer based on this measurement scheme is shown in Fig. 13. One end of the twin beams is anchored while the other end is attached on two flexural suspension beams. A plate electrode is placed aside the clamped end of the twin beams to form a parallel-plate gate capacitor for storing charge. The charge is applied across the gate capacitor thus an electrostatic force is induced and exerted along the axial direction of the beams to lead to the change of elastic properties. Symmetric parallel-plate capacitors are also used to actuate the beams and monitor the motional current induced from resonant vibration.

An energy method is used to analyze how the input charge changes the stiffness of the vibrating beams. The resonant motion by the actuation of external force is described by kinetic energy. Given that the capacitive resonators are usually actuated by harmonic force of an AC voltage, the resonant motion of the beams is assumed to be harmonic and is given by $\Lambda(t) = X_A \sin(2\pi f_R t + \varphi)$, where $\Lambda(t)$ is the time dependent generalized modal coordinate that describes the motion at each point along the beam; X_A is the peak amplitude at the resonant frequency f_R; t is the time and φ is the phase. Then, the modal deformation at a specific vibration mode should be defined. In this case, each beam is double clamped thus the mode shape is given by a position dependent function

$$\Phi_i(x) = cos\beta_i x - cosh\beta_i x + \frac{cos\beta_i L - cosh\beta_i L}{sin\beta_i L - sinh\beta_i L}(sin\beta_i x - sinh\beta_i x),$$
$$\beta_i L = (i + 1/2)\pi \tag{11}$$

where x is the position along the length of the beam L. i represents mode order of the resonator and β is the constant that determines the deformation shape of each mode. Given the resonator is operated in the first order mode of which β_1 is calculated to be 4.73. Next, the motion of the beam can be modeled as $\Omega_1(t, x) = \Lambda(t) \times \Phi_1(x)$. Thus, the total potential energy U_P consists of the bending energy U_{PB} and the axial strain energy U_{PS} produced by the pre-applied tension or compression. U_P can be derived by integrating constituent energies over the length of the beam

$$U_P = U_{P_B} + U_{P_S} = \frac{1}{2}\int_0^L EI\left(\frac{\partial^2 \Omega_1}{\partial x^2}\right)^2 dx + \frac{1}{2}\int_0^L F_C(\frac{\partial \Omega_1}{\partial x})^2 dx \tag{12}$$

where E is the elastic modulus. I is the moment of inertia of the beam and is given by $I = hw^3/12$ in terms of the constant rectangle cross-section, where h and w are the thickness and width of the beam, respectively. F_C is the electrostatic force across the gate capacitor induced by input charge and is determined by $F_C = q^2/2\varepsilon S_G$, q is the input charge. ε is the dielectric coefficient and S_G is the plate area of the gate capacitor. As seen from Eq. (12), the total potential energy is increased while the beam is in tensile state and is decreased for compressive state of the beam. The total kinetic energy U_K can be modeled by integrating the individual kinetic energy at each point of the beam over the length and then plus with the kinetic energy from the motion of the attached masses

$$U_K = \frac{1}{2}\int_0^L \rho A\left(\frac{\partial \Omega_1}{\partial t}\right)^2 dx + \sum_j \frac{1}{2} m_j \left(\frac{\partial \Omega_1(x_j)}{\partial t}\right)^2 \tag{13}$$

where ρ is the material density and A is the area of the cross-section of the beam. m_j is the attached mass on the position x_j of the beam. The equation of motion in the form of potential and kinetic energies according to Lagrange's method is given by

$$\frac{d}{dt}\left(\frac{\partial U_P}{\partial \dot{\Lambda}}\right) - \frac{\partial U_P}{\partial \Lambda} + \frac{\partial U_K}{\partial \Lambda} = 0, \tag{14}$$

$\Lambda(t)$ is position independent and $\Phi_1(x)$ is time independent. Substituting Eqs. (12) and (13), Eq. (14) can be solved as $m_{eff}\partial^2\Lambda/\partial t^2 + k_{eff}\Lambda = 0$. This is exactly the form of Eq. (2), so that the solution of the resonant frequency is given by $f_R = \sqrt{k_{eff}/m_{eff}}/2\pi$. The effective mass m_{eff} and effective stiffness k_{eff} of the resonator are determined by

$$k_{eff} = \frac{d^2U_P}{d\Lambda^2} = \int_0^L EI\left(\frac{\partial^2 \Phi_1}{\partial x^2}\right)^2 dx + \int_0^L F_C\left(\frac{\partial \Phi_1}{\partial x}\right)^2 dx$$

$$m_{eff} = \frac{d^2U_K}{d(\partial\Lambda/\partial t)^2} = \int_0^L \rho A \Phi_1^2 dx + \sum_j m_j \Phi_1(x_j) \tag{15}$$

Substitutions of $\Phi_1(x)$ and F_C into Eq. (15) gives

$$k_{eff} = k + \Delta k \approx \frac{192EI}{L^3} + \frac{16q^2}{5\varepsilon L S_G}$$

$$m_{eff} = m_B + m_A \approx 0.375\rho AL + \sum_j m_j \Phi_1(x_j) \tag{16}$$

where m_B and m_A are the mass of the beam and the attached structure. For this structure of gate capacitor, an attractive force is induced by input charge and applied on the beams. The tensile state induced by the input charge can be represented by stiffness modulation Δk on original mechanical stiffness k. Accordingly, the resonant frequency f_R can be rewritten as

$$f_R = 2\pi\sqrt{\frac{k_{eff}}{m_{eff}}} = f_0\sqrt{1 + \frac{\Delta k}{k}},\ f_0 = 2\pi\sqrt{\frac{k}{m_{eff}}} \tag{17}$$

f_0 is the natural frequency in the absence of input charge. Since the $\Delta k << k$, Eq. (17) can be expanded into Taylor series, based on that the relative sensitivity Γ_A of the resonant electrometer can be deduced as

$$f_R - f_0 = f_0 \frac{3L^2 F_C}{40Ehw^3}, \ \Gamma_A = \frac{f_R - f_0}{f_0 q^2} = \frac{3L^2}{80\varepsilon E S_G h w^3} \tag{18}$$

In Eq. (18), the situation that the twin beam on which the axial force is equally distributed has been considered. The frequency shift has quadratic relation to the charge being measured. The sensitivity model provides strategies for parametrically designing the resonator to achieve better sensitive nature while ensuring the structure robustness, e.g. properly increasing length to width ratio *L*/*w* or length to thickness ratio *L*/*h*, and reducing plate area S_G of the gate capacitor.

To further increase the sensitivity, one way is to employ mechanical force amplifier between the gate capacitor and the beams, e.g. the micro-leverages. A simple resonator design incorporating dual micro leverages is shown in Fig. 14a. The single stage micro leverage consists of force input and output ports, lever beam and pivot, as shown in Fig. 14b. Constrained by micromachining techniques, the micro pivot is quite different from macro counterparts that use hinges. Instead, the coplanar flexures are commonly used. According to relative position between those components, there are two sorts of leverages classified by whether the output of force amplification is inverted or noninverted. The noninverted type leverage keeps the input and output forces with consistent force direction. Differently, the inverted type leverage can change the force direction of the input. For example, in Fig. 14c, while the input and output ports are on opposite sides of the lever beam with a pivot placed between them, the input attractive force in the gate capacitor will be transformed to a compressive force outputting on the axial direction of the beam. With inverted output from leverage, the original tensile force at the gate capacitor is converted and

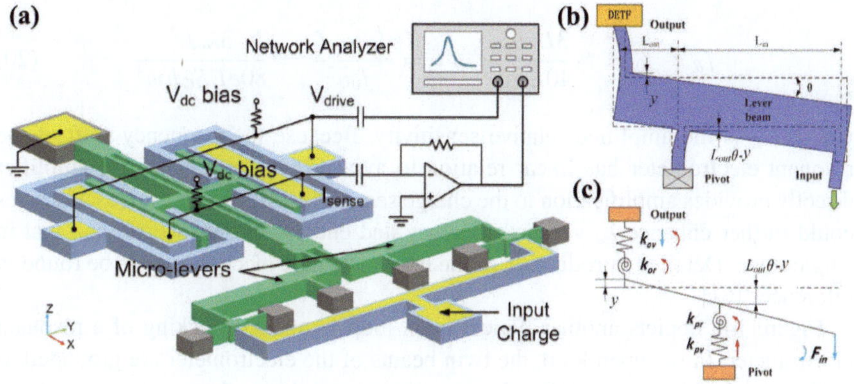

Fig. 14 **a** Schematic of a simple twin beams resonator incorporating dual micro leverages for electrostatic force amplification. **b** Mechanical components of the micro leverage. **c** Mechanic model of the leverage for force amplification analysis

amplified to a compressive force. The differential scheme, which exploits opposite amplification directions of the inverted and noninverted leverages in two resonators for doubling sensitivity and overcoming environmental noises, is frequently used in resonant electrometers [57, 58] and other resonant sensors [59].

To analyze force amplification in electrometry, the lever beam is assumed to be rigid and the residual force due to the bending of the beam during vibration is ignored. The horizontal displacement cancels each other out since the dual micro-levers are distributed symmetrically. The amplification factor λ can be described by the ratio of output force F_{out} to the input force F_{in}, which is given by [60]

$$\lambda = \frac{F_{out}}{F_{in}} = \frac{1/k_{pv}(k_{pr} + k_{or}) - L_{out}L_{in}}{(1/k_{pv} + 1/k_{ov})(k_{pr} + k_{or}) + L_{out}^2} \tag{19}$$

where L_{out} and L_{in} represent the distance between output and input ports to pivot, respectively. k_{pv} and k_{ov} are the elastic spring constants of the suspension beams at the pivot and output port. k_{pr} and k_{or} are the rotational spring constants of the suspension beams at pivot and output port. In an ideal leverage system, the elastic springs are not extensional so that k_{pv} and k_{ov} tend to be infinity. The rotational springs should act like hinges that k_{pr} and k_{or} tend to be zero. Thus, the nominal value of amplification factor approaches $\lambda = -L_{in}/L_{out}$, where the negative sign indicates the inverted force direction at the output port. However, the suspension beam based pivots have nonzero rotational springs, which cause energy loss. Therefore, the effective amplification factor λ_e is less than the nominal one. This energy loss is related to the dimension of the pivots and the initial design of λ. For example, a thinner suspension beam will lead to a larger λ_e/λ. And either increase of L_{in} or decrease of L_{out} will have a lower λ_e/λ. In other words, the energy loss is increased and the efficiency of the leverage is reduced despite increase of λ_e and λ. Overall, λ_e is mainly determined by geometric design of the leverage. While the micromechanical leverages are introduced to the resonant electrometer, Eq. (18) can also be written in the form of

$$f_R - f_0 = f_0\frac{3L^2F_{out}}{40Ehw^3}, \; \Gamma_E = \frac{f_R - f_0}{f_0q^2} = \frac{3\lambda_e L^2}{80\varepsilon E S_G h w^3} \tag{20}$$

where Γ_E is the amplified relative sensitivity. Because the frequency shift of the resonant electrometer has linear relation to axial force, the mechanical amplifier directly provides amplification to the charge sensitivity. Multi-stage micro leverages could further enhance λ_e yet the efficiency and energy loss are worse than that in single stage. Detailed introduction of the micro leverage mechanism can be found in references [61].

Figure 14a depicts another typical open loop frequency tracking of a resonant electrometer. In the open loop, the twin beams of the electrometer are grounded as well as the dual leverages. DC voltages are used to polarize the symmetric parallel-plate capacitors where the motional current will be induced due to periodic change of the gap distance during vibration. A network analyzer is used to provide frequency sweeping on the electrometer by outputting an AC voltage to force the twin beams to vibrate in the vicinity of its natural frequency. The induced current is converted to voltage by a transimpedance amplifier and then is fed back to the network analyzer. The open loop implementation locates the resonant frequency by searching maximum power transmission of the resonator in frequency domain. The other way is to analyze its impedance level in frequency domain to find out peak amplitude thus resonant frequency [62]. Although the twin beams have in-phase and out-of-phase modes with close natural frequencies [60], the symmetrical actuation topology ensures the suppression of the spurious in-phase mode. The out-of-phase vibration mode, as seen in Fig. 13b generally has a higher quality factor due to mutual cancellation of opposing stress waves in the coupling-end of the twin beams. That is the reason why the twin beam resonators are preferred to the single beam resonators in resonant sensors.

For the axial strain modulation charge sensing scheme with the same design to the electrometer in Fig. 14a, the sensitivity is 0.01 Hz/fC^2. The final design of this type resonant electrometer is depicted in Fig. 15a. The resolution is 0.85 fC (5313 e) subject to 63 ppb frequency fluctuation and a resonant quality factor of 8000 [60]. The open loop results for charge measurement are illustrated in Fig. 15b, where the translation of frequency spectrum can be observed as predicted by theory. For open loop test, the entire spectrum should be mapped to find resonant frequency. Open loop requires large numbers of discrete points for frequency sweeping, while in a large frequency span the discrete points are limited, so a considerable frequency fluctuation is caused. In that case, by embedding the electrometer in a closed oscillation loop, the frequency stability and resolution can be improved, which will be introduced later.

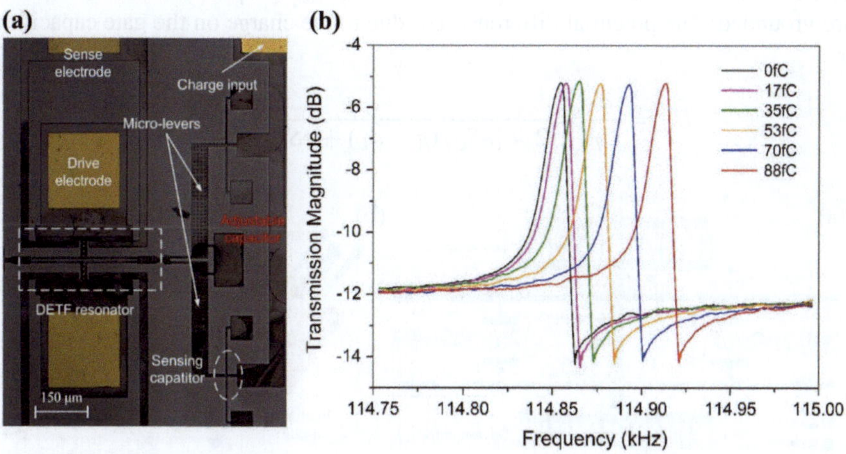

Fig. 15 **a** Optical micrograph of the double ended tuning forks resonator incorporating micro leverages and capacitive transducers. **b** Open loop characterization on spectral response of the resonant electrometer under charge input

3.2 Lateral Stiffness Perturbation

Different from the change of mechanical properties of the beam subject to the axial force, the charge inputted laterally from the direction of vibration can inject energy to the system by accelerating or decelerating the resonant motion, thus modulates the resonant frequency. In some cases, axial strain modulation scheme for force detection is preferred, such as inertial sensors [38]. As for electrometry, lateral stiffness perturbation scheme shows better performance, as will be demonstrated in this section by taking beams resonator as examples.

Figure 16a depicts the mechanism of lateral stiffness perturbation schemes based on a simple twin beams resonator. The electrometer consists of two comb capacitors for resonant motion actuation and detection, and side gate capacitors for gating the charge being measured. The twin beams are physically connected by a guide shuttle at the center, aiming to let the resonator be operated at only in-phase mode, as shown in Fig. 16b. This strongly mechanical coupling will suppress other spurious in plane flexural modes, e.g. out of phase mode. The in-phase mode in this device is designed for implementing lateral stiffness perturbation. The two gate capacitors are placed on the center of the half beam and are formed by sandwiching a bar with two parallel plates.

Mechanical stiffness of this comb drive resonator can be derived by using the energy method aforementioned, as given by $k_B = 48EI/L^3$, where L denotes the length of the half beam. Considering the sandwiched gate capacitor as a whole, the total gate capacitance remains constant due to the small and antisymmetric changes of the two gaps subject to the in-phase mode of the beams. Charge redistributes on the bars due to the change in the gaps during resonance, and keeps an equilibrium potential difference across the two parallel connected gate capacitors. The twin beams are grounded. The potential difference U_C due to the charge on the gate capacitors is

$$U_C = \frac{q}{C_T} = \frac{q}{2 \times [\varepsilon S_G/(g - d_L) + \varepsilon S_G/(g + d_H)]} \tag{21}$$

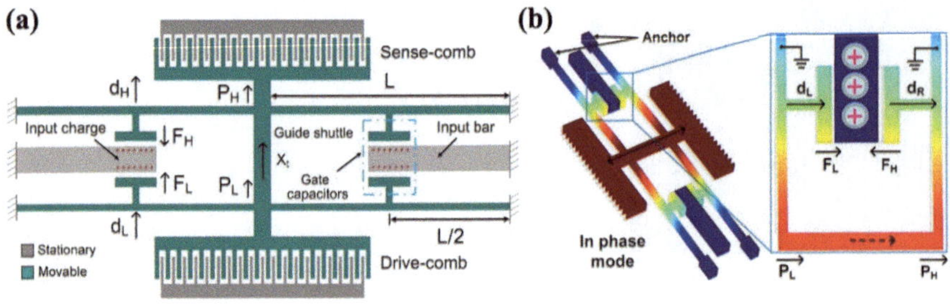

Fig. 16 **a** Lateral stiffness perturbation electrometry scheme based on a simple twin beams resonator. **b** In-phase mode of the beams resonator. The inset shows the zoom in figure of the sandwiched gate capacitor at vibration

where C_T is the total capacitances of the gate capacitors. S_G and g are the plate-area and gap distance of the gate capacitors, respectively. d_L and d_H are the displacement at the midpoint of the underneath and upside parallel plates of the gate capacitor. Therefore, at those midpoints, the charge induced electrostatic forces F_L and F_H can be expressed as

$$F_L = \frac{\varepsilon S_G U_C^2}{2 \times (g - d_L)^2}, \; F_H = \frac{\varepsilon S_G U_C^2}{2 \times (g + d_H)^2} \tag{22}$$

Then, Euler–Bernoulli beam equations [63] and Castigliano's theorem [64] are used to solve the small displacement at those midpoints. Both the small displacements are derived as

$$d_L = \frac{F_L L^3}{192EI} + \frac{x_t}{2}, \; d_H = \frac{-F_H L^3}{192EI} + \frac{x_t}{2} \tag{23}$$

where x_t is the displacement at the center of the beam. Taking account of the mechanical stiffness and the electrostatic force at the midpoint of the half beams, the reaction forces P_L and P_H at the center of the underneath and upside beams are determined by

$$P_L = \frac{12EIx_t}{L^3} - \frac{F_L}{2}, \; P_H = \frac{12EIx_t}{L^3} + \frac{F_H}{2} \tag{24}$$

With the known reaction force and displacement at the center of the beam, the effective stiffness k_{eff} of the twin beam can be deduced. The above analysis is based on the half of the beam, therefore the contribution from the other half should be included while the total effective stiffness is considered. Combining the above equations, a closed form solution is obtained

$$k_{eff} = \frac{2 \times (P_L + P_H)}{x_t} \approx \frac{48EI}{L^3} - \frac{q^2}{16\varepsilon S_G g} \tag{25}$$

The first term on the right-hand side of Eq. (25) equals to the mechanical stiffness of the twin beams. Therefore, the second term represents the effectively perturbed stiffness in terms of charge addition to the gate capacitors. Equation (25) indicates the lateral perturbation by charge is also based on stiffness modulation, but it does not stiffen or soften the beam as axial force. More essentially, it adds a negative electrical spring coupling to transverse mechanical spring of the beam. Substituting k_{eff} to Eq. (17), the relative sensitivity of laterally stiffness perturbation scheme can be derived by

$$f_R - f_0 = f_0 \frac{q^2 L^3}{128E\varepsilon S_G g h w^3}, \; \Gamma_L = \frac{f_R - f_0}{f_0 q^2} = \frac{L^3}{128E\varepsilon S_G g h w^3} \tag{26}$$

Notice that the lateral stiffness perturbation scheme also has quadratic sensitivity, which is consistent with the axial strain modulation counterpart, but the sensitivity depends on the gap of the gate capacitor. By comparing the relative sensitivities of the two measurement schemes, it can be found that

$$\Gamma_L = \frac{L^3}{128E\varepsilon S_G g h w^3} = \Gamma_A \times \frac{80}{3 \times 128} \times \frac{L}{g} \tag{27}$$

The relative sensitivity is mainly determined by the dimension of the beams and gate capacitors. Under the same dimension of that, lateral stiffness perturbation scheme has higher sensitivity, because the length of the beam L is quite larger than the gap distance g in typical designs of the beam resonators and sensors. Actually, the value of g is designed as small as possible even approaching fabrication limit, which is beneficial to electromechanical coupling of the capacitive devices. Therefore, the sensitivity improvement over an order of magnitude is ensured. Further narrowing down the capacitance of the gate capacitor can boost charge sensitivity.

With curved beam resonators, electrometry based on lateral stiffness perturbation has better performance than the one based on axial strain modulation. Figure 17a depicts the curving process of a straight beam to circular beam while the length and width are kept constant. Based on that, Fig. 17b presents a series of resonator designs incorporating curved beams with different curvature from 0 to 180°. One end of each resonator is set free by attaching at the center of a suspension beam

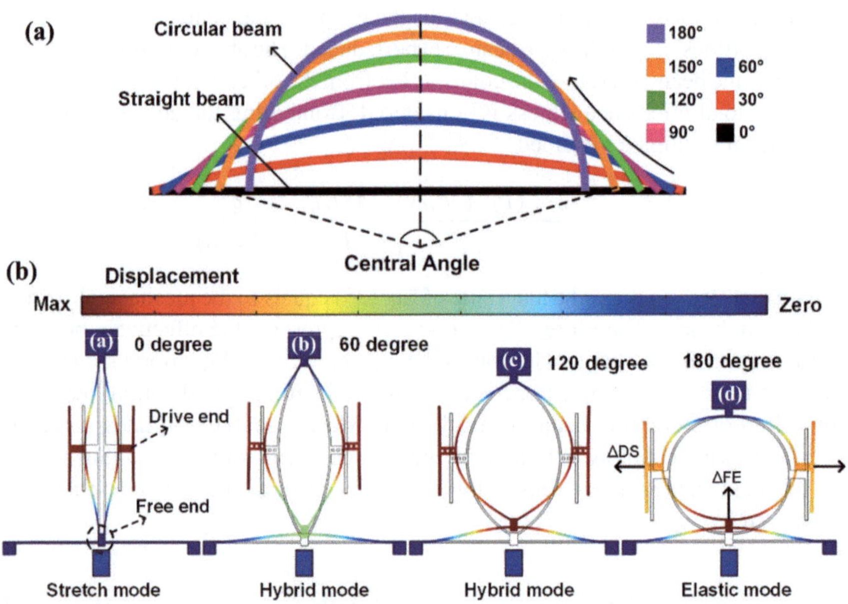

Fig. 17 **a** The curving process of straight beam to circular beam while the length and width of the beam is kept constant. **b** The vibration modes of the curved beam resonator with different curvatures

which has a large length to width ratio to ensure the degree of freedom along axial direction while the other end is anchored. The vibration modes of those resonators are analyzed via finite element method. As observed, the free end of the straight beam resonator is inactive under the stretch mode vibration. In the curved beam resonators, the free end has actual displacement at resonance. In fact, the displacement ratio of the free end and drive end increases with the increase of the curvature [65]. When the beam bends with a curvature from straight (0°) to circular (180°), the vibration mode is changed from stretch mode to stretch/elastic hybrid modes and ultimately reaches elastic mode. By placing a gate capacitor on the free end, the curved beam with curvature between 0 and 180° actually encounters with a mix of axial strain modulation and lateral stiffness perturbation simultaneously while charge loaded. As for the curvature of 180°, the circular beam has no axial direction so that only the lateral perturbation is considered. Therefore, while the beam is curved from straight to circular, the measurement scheme transits from axial strain modulation to lateral stiffness perturbation.

In an attempt to compare the independence and the mix of two charge measurement schemes, a series of finite element simulations are carried out in an established COMSOL multi-physics field. A two-step solver including prestressed analysis and frequency domain analysis has been conducted in the selected electro-mechanical filed. The charge input process is implemented in the prestressed step and then the resonant frequencies are found in the second step. Figure 18 plots the results of the relative sensitivity curves of the resonant electrometer incorporating curved beam resonator with curvature from 0 to 180° and the same beam dimensions. The mixed measurement scheme in hybrid mode curved beam resonators exhibits quadratic nature of charge sensitivity, as the same with the two independent schemes. The results clearly show an increase of sensitivity in the resonant electrometers with highly curved beam. The relative sensitivity reaches the highest value at the beam resonator with a curvature of 180°. A charge sensitivity enhancement over 50 times is achieved in the transition of measurement scheme by manipulating the curvature

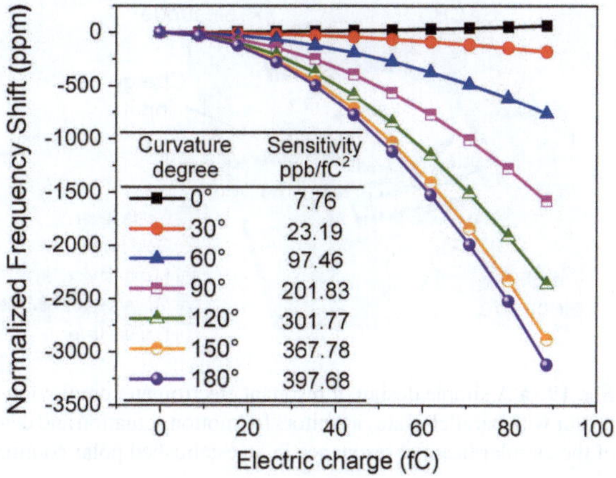

Fig. 18 Finite elements simulation results of the charge sensitivity of mixed and independent measurement schemes in the resonant electrometers based on the curved beam with different curvatures

of the beam. The performances of the lateral stiffness perturbation charge sensing scheme will be demonstrated in the next section.

3.3 Prototype Instantiation

In this section, the practical implementation of resonant electrometry incorporating frequency modulation oscillator will be demonstrated based on the analysis of the two charge measurement paradigms [66]. In the previous analysis, the lateral stiffness perturbation scheme on a circular beam resonator has the best performance. Thus, a simple design of the resonant electrometer employing circular beam resonator is depicted in Fig. 19a. The vibratory element of the electrometer consists of a singly anchored circular beam with the other one end set free. The gate capacitor is formed by the free end of the circular beam and the charge input gate electrode. When small quantities of charge are applied on the gate electrode, the vibrating circular beam will experience an electrostatic attractive force and then change frequency passively in response to this force. Symmetric parallel-plate capacitors are designed for driving the resonator into resonance. The resonance motion is monitored by detecting the induced motional current on the sensing electrodes.

As a result of the spurious mode suppression by the symmetric driving topology, the mechanical resonator is operated at an elastic mode deformed with elliptical shape, as depicted in Fig. 19b. The blue and red dash lines illustrate the deformation of the elliptical mode shape at resonance. z and y axes denote the resonator actuation and charge perturbation directions, respectively, corresponding to the case in Fig. 19a. Figure 20 illustrates two operation topologies for the circular beam resonator, which are open loop implementation, and closed loop for stable oscillation by embedding the resonator in a phase locked loop, as theoretically analyzed above. The main

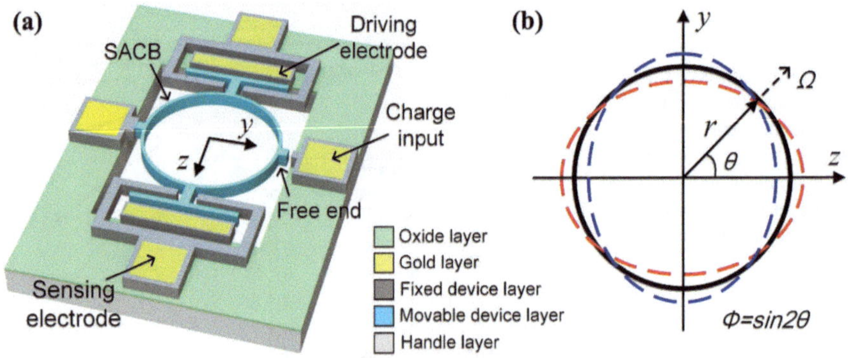

Fig. 19 **a** A simple design of resonant electrometer employing single anchored circular beam resonator with parallel-plate capacitors for motion actuation and detection. **b** The elliptical deformation of the circular beam at resonance in an established polar coordinate system

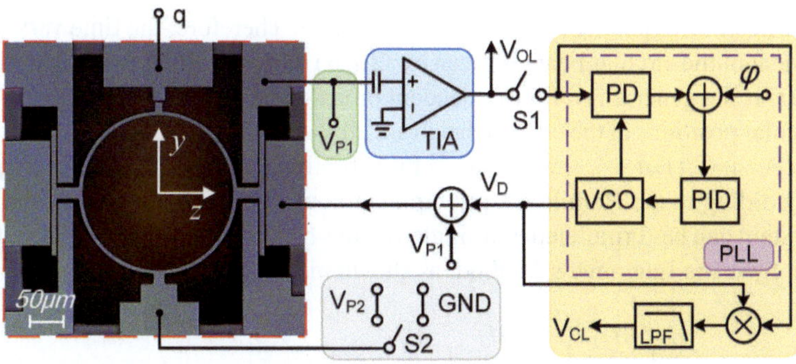

Fig. 20 Optical micrograph of the mechanical electrometer and the experimental implementation for open loop and closed loop operations. The output current from the sense port passes through a trans-impedance amplifier (TIA) for conversion as a voltage V_{OL} which is the output in open loop while S1 is switched off. To close the loop, we switch S1 on to feedback VOL to a phase locked loop (PLL). Within PLL, the phase of V_{OL} is detected by a phase detector (PD) for feeding to a proportional-integral-derivative (PID) controller to enable control on voltage controlled oscillator (VCO) to output V_D with shifted phase by φ. Thus, we precisely create phase offset conditions between V_{OL} at output port and V_D at input port for satisfying the Barkhausen criteria and building oscillations. The sustained oscillation signal is demodulated for output (V_{CL}) in real-time. LPF: low pass filter; GND: ground

body of the circular beam is grounded by switching $S2$ to GND and the resonator is polarized at the driving and sensing electrodes by enabling $V_{P1} = 20$ V. The electrostatic actuation is used, as shown in Fig. 20. An AC voltage V_D ($V_D << V_{P1}$) is applied to provide dynamic force for driving the resonator in the vicinity of natural frequency, while a DC bias voltage V_{P1} is applied for polarizing the parallel-plate capacitors for monitoring resonant motion. With a transimpedance amplifier (TIA), the displacement induced motional current at the sensing electrodes is amplified as V_{OL} which is the output at open loop while $S1$ is switched off. While $S1$ is switched on, the resonance loop is closed up by with feedback of V_{OL} to a phase locked loop. The phase difference between V_{OL} and the driving AC voltage V_D can be precisely controlled with PLL. Specific offset phase conditions are created between the input and output ports of the oscillator. Thus the Barkhausen's criteria are satisfied and a stable closed loop oscillation can be realized to track resonant frequency in real time. Then the oscillation amplitude and frequency are demodulated and outputted as V_{CL}. Moreover, the resonant electrometer is placed in a vacuum chamber with pressure level under 20 mTorr at room temperature.

Similarly, the circular beam resonator is considered as an undamped spring-mass-damper vibration system that can store and convert energy from external input, including the energy from the capacitive force of driving voltage and perturbation force of input charge. A polar coordinate system (r, θ) is established, as shown in Fig. 19b, where r is the radius and θ is the polar angle. In the case of a thin circular beam in which the mid-surface radius is much larger than the radial width, the elliptical mode shape becomes independent of the mid-surface radius and can be given by an angular position dependent function $\Phi(\theta) = \sin 2\theta$. Therefore, the time varying displacement of the circular beam is given by $\Omega(t, \theta) = \Lambda(t) \times \Phi(\theta)$, where t is the time and $\Lambda(t)$ is the generalized time dependent modal coordinate. The motions at different angular positions of the circular beam are all assumed to be harmonic and given by $\Lambda(t) = A_R \sin(2\pi f_N t + \varphi)$, where A_R is the vibration amplitude at natural frequency f_N. According to the Hamilton's principle without external perturbation, the vibration system can be formulated as a single-degree-of-freedom differential equation of motion by using Lagrange's Method, as shown in the following equations

$$\frac{d}{dt}\left(\frac{\partial U_P}{\partial \dot{\Lambda}}\right) - \frac{\partial U_P}{\partial \Lambda} + \frac{\partial U_K}{\partial \Lambda} = 0$$
$$U_K = \frac{EI}{2R^3}\int_0^{2\pi}\left(\Omega + \frac{\partial^2 \Omega}{\partial \theta^2}\right)^2 d\theta$$
$$U_P = \int_0^{2\pi}\frac{1}{2}\rho w h\left(\frac{\partial \Omega}{\partial t}\right)^2 d\theta + \sum_j \frac{1}{2} m_j \left(\frac{\partial \Omega(\theta_j)}{\partial t}\right)^2 \tag{28}$$

where U_P and U_K are the potential energy and kinetic energy of the vibration system. E, I and ρ are the elastic modulus, moment of inertia and density, respectively. R, w and h are the mid-surface radius, radial width, thickness of the circular beam, respectively. m_j denotes the attached mass on the beam at different angular positions θ_j. Given that $\Phi(\theta)$ is time independent and $\Lambda(t)$ is position independent, Eq. (28) can be further deduced as

$$m_R \ddot{\Lambda} + k_R \Lambda = 0$$
$$k_R = \frac{d^2 U_P}{d\Lambda^2} = \frac{EI}{2R^3}\int_0^{2\pi}\left(\Phi_2 + \frac{\partial^2 \Phi_2}{\partial \theta^2}\right)^2 d\theta$$
$$m_R = \frac{d^2 U_K}{d(\partial \Lambda/\partial t)^2} = \int_0^{2\pi} \rho w h \Phi_2^2 d\theta + \sum_j m_j \Phi_2(\theta_j)^2 \tag{29}$$

where k_R and m_R are the mechanical stiffness and effective mass of the circular beam. Substituting the harmonic motion term $\Lambda(t)$ to Eq. (29), the mechanical natural frequency f_N of the resonator can be deduced as $f_N = \sqrt{k_R/m_R}/2\pi$. As for the electrostatic motion actuation and detection, an electrical stiffness k_E is introduced to the system [48], given by $k_E = -4\varepsilon S_P V_{P1}^2/g^3$, where ε is the dielectric constant; S_P and g are the plate area and gap distance of the parallel-plate capacitors. It is assumed that the vibration amplitude A_R of the elliptical mode shape along z direction is much smaller comparing to the gap distance of the gate capacitor. When input charge is loaded on the gate capacitor, the energy addition U_C to the vibration system is

$$U_C = \frac{G^2}{2\varepsilon S_C} \int_{-A_R}^{A_R} \frac{q^2}{(G+z)^2} dx, k_C(q) = \frac{dU_C}{dA} = -\frac{q^2}{\varepsilon S_C G} \tag{30}$$

where $k_C(q)$ denotes the effectively perturbed stiffness of the resonator deduced from the derivative of the added energy U_C. S_C is the plate area of the gate capacitor. Equation (30) indicates that the stiffness perturbation induced by charge perturbation is independent of vibration amplitude. Combining all the equations above, the frequency response $f_C(q)$ of the circular beam resonator subject to the capacitively actuation and charge perturbation is derived

$$f_C(q) = \frac{1}{2\pi}\sqrt{\frac{k_R + k_E + k_C(q)}{m_R}} = f_R\sqrt{\frac{k_C(q)}{k_R + k_E}} \tag{31}$$

where f_R is the resonant frequency under forced vibration and zero charge input. Expanding Eq. (31) into Taylor series, the fractional frequency change γ is given by

$$\gamma = \frac{f_C(q) - f_R}{f_R} = -\frac{q^2}{2\varepsilon S_C G(k_R + k_E)} \tag{32}$$

Therefore, the relative sensitivity Γ has a quadratic nature and is calculated to be $\Gamma = \gamma/q^2 = -0.25$ ppm/fC^2 using the design parameters given in Table 1 while V_{P1} is fixed at 20 V. Equation (32) concludes fundamental physics of the charge measurement by lateral stiffness perturbation scheme.

Figure 21 shows the frequency spectra of the resonant electrometer by means of open loop and closed loop characterization under 0 fC quantities of charge inputted while V_D is fixed at 10 mV. The dot plots are the results from closed loop mapping while the solid and dash line plots are the open loop results from frequency forward and backward sweeping of a frequency response analyzer, respectively. In the open loop test, the spectra under forward and backward frequency sweeping are consistent, which indicates the device is operated in linear regime. The resonant frequency is found to be 143.029 kHz under the elliptical mode shape, and is only 0.4% deviation from the theoretical result of 142.455 kHz calculated from Eq. (31). In the closed loop, the phase of the loop is locked at different values. The corresponding spectra, including oscillation frequency and amplitude, are mapped, which coincide with

Table 1 Design parameters of the resonant electrometer

Design parameters (symbols)	Value (units)
Mid-surface radius R	127.3 μm
Radial width w	5 μm
Beam's thickness h	25 μm
Plate area of the gate capacitor S_G	300 μm^2
Gap distance of the gate capacitor G	2.5 μm

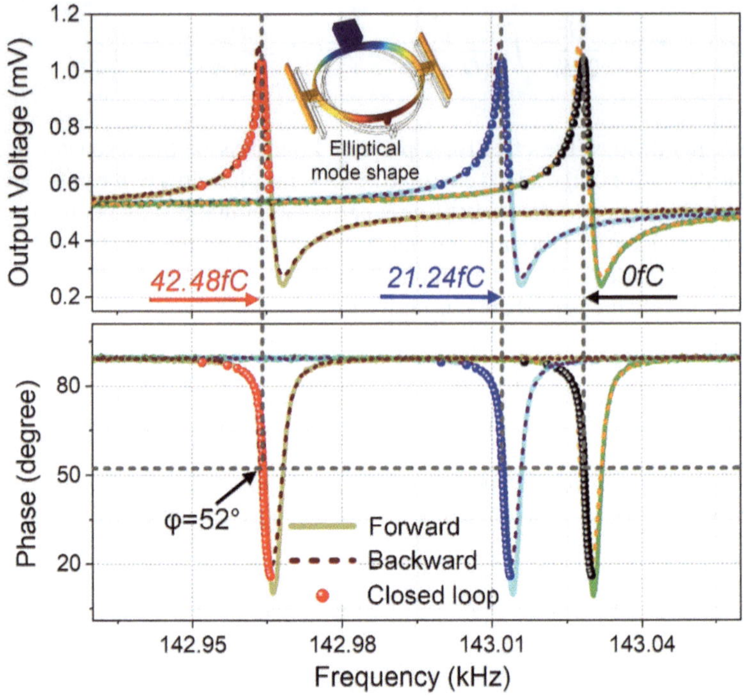

Fig. 21 Open loop and closed loop characterizations of the spectral response of the resonant electrometer under different quantities of charge applied on the gate capacitor. The inset shows the elliptical mode shape of the mechanical resonator derived from COMSOL finite element simulation

the results from open loop test. This verifies the effectiveness of frequency tracking functionality of the closed loop implementation. Then with 21.24 and 42.48 fC quantities of charge applied, the resonant frequencies from open loop and closed loop results consistently shift downward by −112.56 and −443.97 ppm with deviation errors less than 1.6%. Especially in the closed loop, while the oscillator is operated at different locked phases and output voltages, the changes in the frequency responses with respect to charge addition is nearly identical. This proves the linearity nature of the frequency modulation within the resonance bandwidth in stiffness perturbation scheme, which agrees with the theoretical analysis in Sects. 2.1 and 2.2. As previously analyzed, the amplitude to frequency and phase to frequency noise conversion will all reach minimal at the resonance peak. Thus, the noise is at the lowest level and a high resolution can be achieved. Therefore, the phase of the loop is preferred to be locked at the resonance peak for oscillation. The phase is $\varphi = 52^\circ$ according to Fig. 21. Therefore, the phase of the loop is kept locked at 52° for the following measurements.

The relative sensitivity of the electrometer at the resonance peak in closed loop is characterized and quadratically fitted to be -0.247 ppm/fC^2, as shown in Fig. 22 and the inset table. The results from open loop test and theoretical models are also plotted

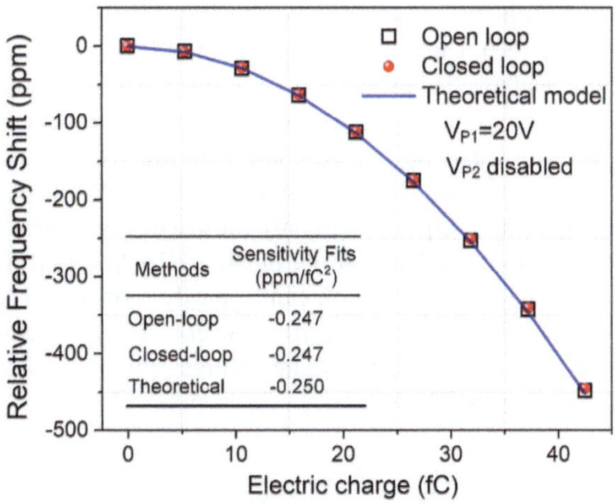

Fig. 22 Quadratic sensitivity curves of the stiffness perturbation charge sensing scheme from theoretical, open loop and closed loop results. The inset table shows the values of the fitted and calculated sensitivities for comparison

and summarized, where a sensitivity deviation is less than 1.2% between theoretical and experimental results. Of particular interest, the resonant frequency of the device has linear relation to the charge induced electrostatic force across the gate capacitor, which yields a quadratic nature of frequency shift in charge measurement. This gives rise to larger frequency shift in response to charge addition at the regime with large charge quantities input, which is beneficial to frequency tracking hence charge resolution. Based on that, a sensitivity amplification scheme for charge measurement based on charge pre-biased states is introduced as follows.

To further carry out the charge pre-biased scheme, the systematic implementation of the electrometer is changed. $S2$ is switched to connect with V_{P2} and meanwhile V_{P1} is disabled so that the electrometer is polarized on the circular beam directly. In this case, the gate capacitor is pre-charged by V_{P2}, leading to a pre-biased charge Q_P on the gate capacitor, given by $Q_P = 2\varepsilon S_C V_P^2/G$. Therefore, by taking this pre-biased state as outset, the new fractional frequency shift in response to charge addition is given by

$$\gamma_P = \frac{f_C(Q_P + q) - f_C(Q_P)}{f_C(Q_P)} = \sqrt{1 + \frac{k_C(Q_P + q) - k_C(Q_P)}{k_R + k_E + k_C(Q_P)}} - 1 \tag{33}$$

According to Eq. (30), it can be obtained that $k_C(Q_P + q) = -(Q_P + q)^2/\varepsilon S_C G$. Substituting that into Eq. (33) and then expanding Eq. (33) into Taylor series with higher order terms neglected, the fractional frequency shift has the form of

$$\gamma_P = \frac{-q^2 + 2qQ_P}{2\varepsilon S_C G(k_R + k_E + k_C(Q_P))} \tag{34}$$

Equation (34) indicates that this charge pre-biased scheme turns the quadratic sensitivity to a parabolic sensitivity by introducing a linear term $2qQ_P$ and is also more sensitive comparing to the original stiffness perturbation scheme described by Eq. (32). Under the configuration of Eq. (32), either positive and negative charge will lead to a downward shift of frequency. Differently, in the pre-biased states, given that Q_P is positive, the charge polarity can be distinguished due to that, the fractional resonant frequency γ_P will shift upward when measuring positive charge and shift downward when measuring negative charge. While $Q_P >> q$, neglecting q^2 term in the numerator of Eq. (34) will only cause a neglectable measurement error. By assuming the error is less than 1‰, we obtain $q^2/2qQ_P <$ 1‰. By solving the relation, we have $Q_P/q > 500$. This relation can be easily satisfied when the quantity of charge being measured is in the range from several to hundreds of electrons. Therefore, under the condition of $Q_P >> q$, the relative sensitivity Γ_P at the pre-biased states can be linearized as follows

$$\Gamma_P = \frac{\gamma_P}{q} = \frac{2Q_P}{2\varepsilon S_C G(k_R + k_E + k_C(Q_P))}, \ (Q_P >> q) \tag{35}$$

This model describes a linearized region in which frequency shifts linearly in response to the input charge. By comparing Eqs. (32) and (35), we obtain $\Gamma_P \approx 2Q_P \times \Gamma$, which indicates in the linearized region, the pre-biased charge Q_P functions as an amplification to the sensitivity of stiffness perturbation scheme in terms of the measurement of charge addition on the gate electrode.

To intuitively illustrate the underlying physics, the quadratic relation between the frequency change and input charge of the stiffness perturbation scheme is plotted in Fig. 23. The sensitivity amplification actually benefits from the sensitive

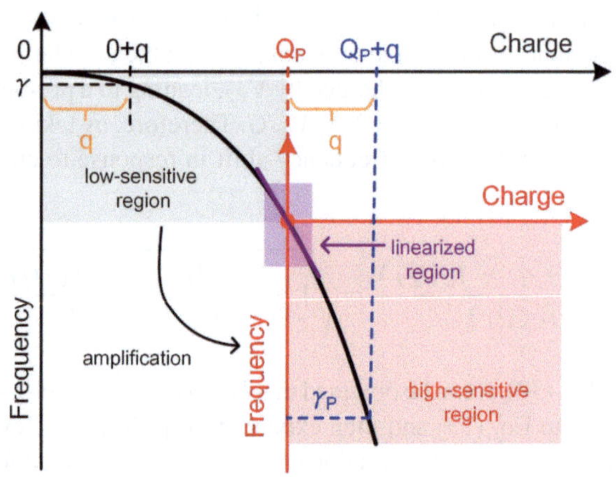

Fig. 23 Illustration of the sensitivity and resolution amplification of the charge pre-biased scheme based on the quadratic sensitivity curve of the stiffness perturbation charge sensing scheme. The coordinate system with black arrow line is for original stiffness perturbation scheme while the inside coordinate system with red arrow line is for the measurement under charge pre-biased states. The amplification scheme introduces charge pre-biased states that move the charge measurement to a more highly sensitive region of the device and a region with linearized sensitivity can be predicted

quadratic nature of stiffness perturbation scheme where dramatic changes in frequency response occur at a region with large quantities of input charge. The strengthening of the charge pre-biased states in the electrometer acts as moving the measurement along the quadratic curve to a much higher sensitive region. Thus, the relative sensitivity is significantly boosted and meanwhile can be linearized using the tangent curve at the outset point of the charge pre-biased regime when measuring charge at electron level. The linearized region expands with the increase of Q_P. The extent of amplification is determined by the fundamental quadratic nature of stiffness perturbation scheme and the quantities of pre-biased charge. According to the theoretical model and designed parameters in Table 1, while 42.48 fC quantities of Q_P is precharged on the gate capacitor, Γ_P can reach a value of 3.5 ppb/e. For high quality factor mechanical oscillators, sub-1 ppb frequency stability is achievable, so that the resonant electrometry based on frequency modulation of mechanical oscillators is possible to approach sub-electron resolution.

As the theoretical analysis above, with known quantities of the pre-biased charge on the gate capacitor, the fractional frequency shift will be amplified and changed parabolically with charge addition. To verify that, $S2$ is switched to V_{P2} to pre-charge the gate capacitor and V_{P1} is disabled. In this case, V_{P2} also provides polarization to the parallel-plate capacitors to enable the output of motional current during resonance. As shown in Fig. 24, the relative sensitivities under four sets of pre-applied

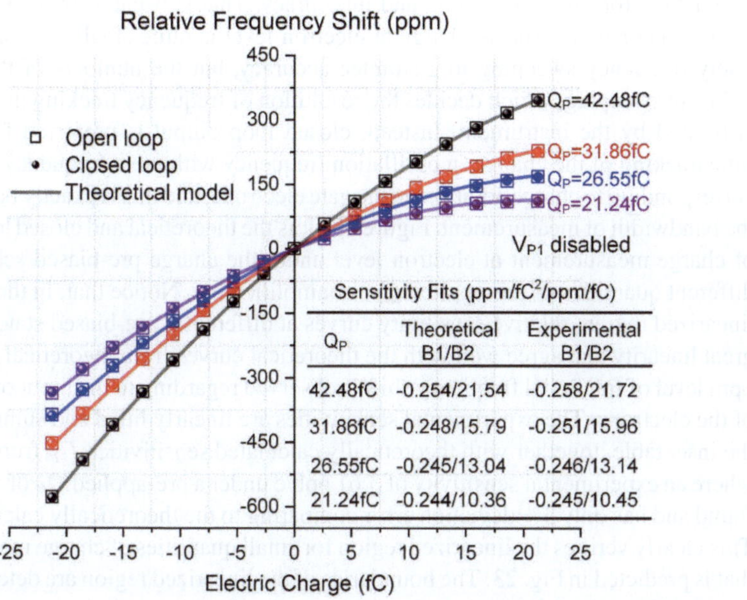

Fig. 24 The parabolic sensitivity curves of the charge pre-biased scheme from theoretical model, open loop and closed loop results under different quantities of pre-applied Q_P. The inset table includes the parabolically fitted sensitivities of closed loop results and the calculated sensitivities from Eq. (34)

Q_P from 21.24 to 42.48 fC are characterized by means of open loop and closed loop measurement. And those sensitivities all present parabolic form. The experimental results from open loop and closed loop tests all agree well with the theoretical predictions from Eq. (34). We further use $\gamma_P = B1 \times q^2 + B2 \times q$ to fit the experimental parabolic curves form closed loop results, as shown in the inset table of Fig. 24, together with the calculated coefficients of $B2 = -2Q_P \times B1$ and $B1 = -1/[2\varepsilon S_C G(k_R + k_E + k_C(Q_P))]$ from Eq. (34). The maximum deviation error between calculated and experimental results is less than 1%. At different quantities of Q_P, the quadratic term $B1$ have values close to the quadratic sensitivity in Fig. 22. The charge pre-biased scheme mainly introduces an additional linear term $B2$ to lead to the enhancement in sensitivity. In fact, when increasing Q_P by two folds from 21.24 fC, the quadratic term $B1$ is only increased by 5.3% while the linear term $B2$ is boosted by 208%. This proves that the pre-applied Q_P enhances the quadratic nature of charge measurement to an amplified parabolic form. Within the measurement range of $\pm Q_P$, the linear term $B2$ dominates sensitivity. Instead, the quadratic term $B1$ will start to dominate out of that range of $\pm Q_P$. Meanwhile, the polarity of input charge is distinguished as the frequency shifts upward for measuring positive charge and shifts opposite for measuring negative charge.

Onward to the charge measurement at electron level, the amplification by charge pre-biased scheme will be more significant and can be effectively translated to the enhancement in resolution. However, the method of open loop measurement will be impractical for time-consuming and inaccuracy. That is because frequency shift of the resonator in measuring charge at electron level is quite small and that requires finely frequency sweeping to guarantee accuracy, but the numbers of the discrete points in sweeping, which decides the resolution of frequency tracking in open loop, is limited by the instrument. Instead, closed loop output is preferred for the real time tracking of the change in oscillation frequency with great frequency resolution corresponding to charge addition on the gate electrode, and this is exactly beneficial to the bandwidth of measurement. Figure 25 plots the theoretical and closed loop results of charge measurement at electron level under the charge pre-biased scheme with different quantities of pre-applied Q_P for amplification. Notice that, in the predicted linearized region, relative sensitivity curves at differently pre-biased states all show great linearity and agree well with the theoretical curves from theoretical models. A ppm level of fractional frequency shift is observed regarding to the input of hundreds of the electrons. The experimental sensitivities are linearly fitted and summarized in the inset table, together with theoretically calculated sensitivities Γ_P from Eq. (35), where an experimental sensitivity of 3.61 ppb/e under a pre-applied Q_P of 42.48 fC is found and has only 3% deviation error comparing to the theoretically calculated Γ_P. This clearly verifies the linearized region for small quantities of charge measurement that is predicted in Fig. 23. The boundaries of the linearized region are determined by the amplification of pre-applied Q_P and the nonlinear error that a charge measurement can undergo. For example, for 1% affordable measurement error and a pre-applied Q_P of 42.48 fC, the linear region can be calculated to be ± 5310 e according to Eq. (34). Figure 26 shows the oscillation frequency output in real time in response to the charge input by 67 electrons per step. By frequency average in a certain period,

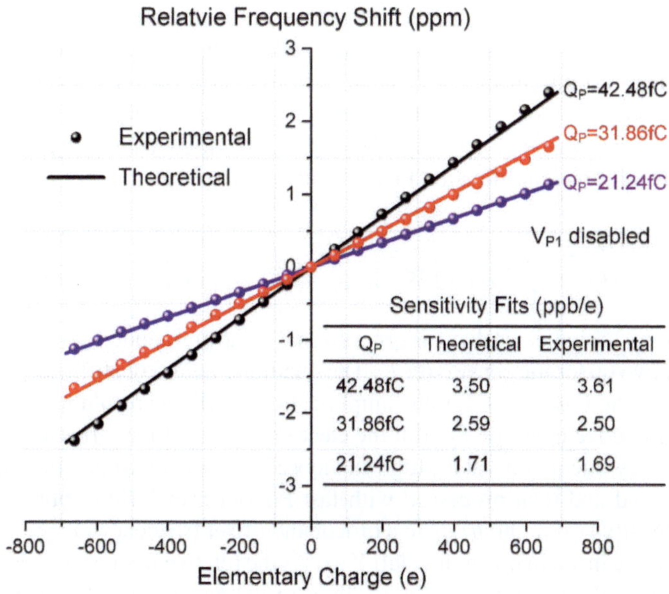

Fig. 25 The linear sensitivity curves of the charge pre-biased scheme from theoretical model and the closed loop results under different quantities of pre-applied Q_P. The inset table shows the linearly fitted sensitivities of closed loop results and the calculated sensitivities from Eq. (35)

Fig. 26 **a** Real time charge detection with step charge input of 67 electrons at a Q_P of 42.48 fC

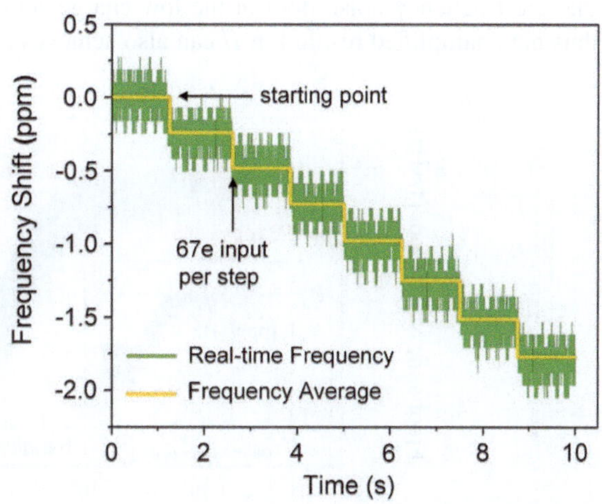

charge detection at electron level is successfully realized. A few electrons can be detected but with smaller frequency change over the averaged random noise. The frequency fluctuation at the start point indicates the fundamental limit of real time charge monitoring, which will be then decomposed in a broadband offset frequency to find charge resolution associated with the loop bandwidth.

In order to analyze the translation of amplification and linearization of the sensitivity into resolution, the relative frequency noise floor of the electrometer is defined as $f(\omega)$. The resolution for the original stiffness perturbation scheme is given by $\Pi = \sqrt{f(\omega)}/|\Gamma|$ while the enhanced resolution for the charge pre-biased scheme in the linearized region is given by $\Pi_P = f(\omega)/|\Gamma_P|$. By combining Eqs. (32) and (35), the resolution amplification factor κ can be derived as

$$\kappa = \frac{\Pi}{\Pi_P} \approx \frac{2Q_P}{\sqrt{f(\omega)}\sqrt{2\varepsilon S_C G(K_B + K_E + K_C(Q_P))}} \approx \frac{2Q_P\sqrt{\Gamma}}{\sqrt{f(\omega)}} \tag{36}$$

Equation (36) indicates that κ is proportional to the pre-applied Q_P and root square of the relative quadratic sensitivity Γ. The quadratic nature of stiffness perturbation scheme lays the foundation for this amplification on charge resolution.

To characterize the noise floor of the electrometer, the time series data of oscillation frequency fluctuations at the high and low charge sensitive operation points have been recorded and then processed with fast Fourier transform to obtain the noise spectral density $f(\omega)_H$ and $f(\omega)_L$ in a broadband offset frequency range, as shown in Fig. 27 and the inset. In this figure, left Y axis is the relative frequency noise floor and right Y axis is the input-referred noise floor rescaling from left Y axis by dividing the linearized sensitivity. In the main figure, the noise characterization is done under a pre-applied charge quantity of 42.48 fC due to the best sensitivity among others. A charge noise floor of 0.17 e/$\sqrt{}$Hz is derived at a white noise level of 0.61 ppb/$\sqrt{}$Hz obtained from the horizontal dash line in Fig. 27. In comparison, the inset shows that the frequency noise floor at the low charge sensitive regime is 1.1 ppb/$\sqrt{}$Hz, thus the unamplified resolution Π can also achieves a level of 417 e/$\sqrt{}$Hz, and that

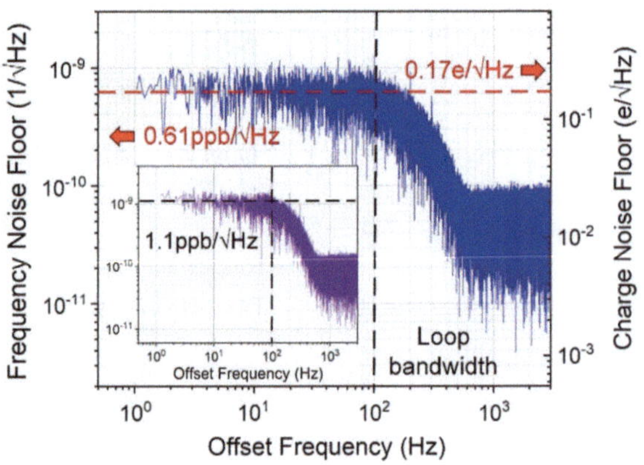

Fig. 27 Measured frequency noise floor as a function of offset frequency at charge pre-biased state with Q_P of 42.48fC (main figure) and the low sensitive regime without charge pre-bias (inset). The left Y axis is the relative frequency noise floor and the right Y axis is the input charge referred noise floor rescaling from left Y axis by dividing the linearized sensitivity to intuitively show the resolution of the resonant electrometer

is benefited from the highly sensitive nature of the stiffness perturbation scheme. The experimental amplification factor κ is calculated to be 2453 which is very close to the estimation of 2333 from Eq. (36). Experimental noise results prove that the frequency noise floor does not scale with the charge sensitivities of the operation points in resonant electrometry. The metrology of resonant electrometry can effectively decouple charge sensing from other noise sources in frequency domain. Thus, $f(\omega)_H$ and $f(\omega)_L$ mainly come from thermal mechanical noise of the vibration and electronic noise of the loop. $f(\omega)_H$ is slightly lower than and $f(\omega)_L$ because that the charge pre-biased states increase the driving force on the transducers and the subsquent increase of the mechanical motion is beneficial for the phase noise of the oscillator hence the frequency noise [52, 53]. The independence of frequency noise with respect to charge sensitivity ensures effective translation of sensitivity amplification to the resolution. The upper boundary of the resonant electrometer is limited by the pull-in effect in the gate capacitor during the input of large quantities of charge. Pull-in effect is a well-known phenomenon in miniaturized capacitive transducers that limits static and motional mechanical displacements. For the gate capacitor with capacitance of 1.062fF, the upper limit of charge detection is found as 770fC, which yields a dynamic range over 149dB. The frequency noises at the region of high offset frequencies are significantly attenuated because of the use of low pass filter at the output of the closed loop. The noises presented in Fig. 27 are a mix of mechanical noises and electronic noises. The dominated mechanical noise is the thermoelastic noise $f(\omega)_M$ from Brownian motion of the mechanical oscillator, which can be estimated by [23]

$$f(\omega)_M \approx \sqrt{\frac{k_B T \Delta f}{2\pi K_B f_N Q A_R^2}} \tag{37}$$

where k_B is the Boltzmann constant and T is the temperature. Δf is the resolution bandwidth. Q is the quality factor of the mechanical oscillator, the value of which is derived as 50,000 by 3-decibel method according to the frequency spectra in Fig. 21. Accordingly, the mechanical noise is estimated to be 1.7 ppt/$\sqrt{}$Hz and this yields a charge noise floor of 4.88×10^{-4} e/$\sqrt{}$Hz. The analysis on mechanical noise indicates that the resolution of the resonant electrometer is mainly dominated by the electronic noises from oscillation enabling circuit. The experimental prototype for implementing resonant electrometry shows extremely sensitive charge measurement with sub-electron resolution. The established theoretical models also provide strategies that can further enhance the resolution of resonant electrometry for several orders of magnitude, e.g. narrowing down the gap of the gate capacitor toward nanometers, strengthening the pre-biased state to let the measurement move into a more sensitive region or properly integrating oscillatory circuit on chip to minimize electronic noises.

Although the scaling law provides many advantages to the miniaturized sensors, such as smaller footprint, greater sensitivity and larger dynamic range, it also gives rise to a higher noise level, particularly for analog signals. This is the root cause that impedes the resolution of the devices with output metrics of analog or amplitude signal. Monitoring on the quasi-nature frequency from a oscillator as sensor output is less affected by noise conversion from vibration amplitude, feedthrough [67] or other analog noise sources. The use of PLL for precisely controlling phases in the loop for oscillation is beneficial to suppress phase noise and thus frequency instability. Therefore, the sensory method may be instructive for developing high-performance mesoscopic oscillators and sensors, including providing alternative perspective for overall loop control strategy and leading to a change in output metric of interest. Additionally, the highly force sensitive mechanical oscillator benefits from the scheme of stiffness perturbation that uses force to perturb modal vibration. This shows a new way for sensing contactless forces, which can be widely adopted for measuring most of the physical quantities that can generate electric, magnetic and inertial forces, such as magnetic and electric fields, mass and pressure, acceleration and angular rate, etc. In that case, the method of stiffness perturbation and closed loop frequency monitoring can be developed as a common mechanism for a variety of sensors, as well as the concept of exploiting nonlinear nature of the sensitivity for internal resolution amplification.

More importantly, the micromachined electrometer can be integrated on-chip as a micromodule for embedding in the scientific instruments that requires high-precision and real-time electrometry. For example, one can use the device as a core component in mass spectrometry system for acquiring ion information to determine elemental and isotopic signatures of a sample, chemical composition and identity of a mixture sample. Besides, charge detection is the basis for measuring most of the electrical parameters. Electric potential, voltage, current are the different existing forms of charge. With proper designs of conversion circuits, the resonant electrometry implementation can be utilized as a key technology to develop advanced instruments, such as voltmeters, ammeters and multimeters, etc.

4 Conclusions

In this chapter, underlying microscopic vibration dynamics for developing the new concept of resonant electrometry at room temperature are introduced with strategies of systematic sensory implementations for practical applications. Capacitive-platform-based micromachining technology enables designers the flexibility to optimizing the dimensional parameters or patterning complex resonating structures for enhancing the sensitivity and resolution. While the dimensions of the vibrational electronics approaching micro and nano scale, oscillational noises become significant. Therefore, currently the best charge resolution by resonant electrometry is at the order of 10^{-1} e/Hz$^{1/2}$. Further reduction on oscillational noises can be realized by incorporating micromechanical oscillators with the concepts of nonlinear dynamics [68], internal resonance [69] and coupling array [70] for frequency and phase stabilization.

Acknowledgements The authors gratefully acknowledge helpful discussions with Ashwin Seshia, Hemin Zhang, Milind Pandit, Guillermo Sobreviela and Jiangkun Sun.

References

1. A. Menzela, A.T.-H. Lin, P. Estrela, P. Li, A.A. Seshia, Biomolecular and electrochemical charge detection by a micromechanical electrometer. Sens. Actuator B **160**, 301 (2011)
2. Y.Y. Wei, J. Weis, K.V. Klitzing et al., Single-electron transistor as an electrometer measuring chemical potential variations. Appl. Phys. Lett. **71**(17), 2514–2516 (1997)
3. A.K. Naik, M.S. Hanay, W.K. Hiebert, X.L. Feng, M. Roukes, Towards single-molecule nanomechanical mass spectrometry. Nat. Nanotechnol. **4**, 445–450 (2009)
4. F. Krueger, J. Larson, Chipmunk IV: development of and experience with a new generation of radiation area monitors for accelerator applications. Nucl. Instrum. Methods Phys. Res. A **495**, 20–28 (2002)
5. J. Lee, N. Tallarida, X. Chen, L. Jensen, V. Apkarian, Microscopy with a single-molecule scanning electrometer. Sci. Adv. **4**, 5472 (2018)
6. A.N. Vamivakas, Y. Zhao, S. Falt, A. Badolato, J.M. Taylor, M. Atature, Nanoscale optical electrometer. Phys. Rev. Lett. **107**, 166802 (2011)
7. A.H. Compton, K.T. Compton, A sensitive modification of the quadrant electrometer: its theory and use. Phys. Rev. **14**, 85 (1919)
8. R.J. Schoelkopf, P. Wahlgren, A.A. Kozhevnikov, P. Delsing, D.E. Prober, The radio-frequency single-electron transistor (RF-SET): a fast and ultrasensitive electrometer. Science **280**, 1238 (1998)
9. Y. Tang, I. Amlani, A.O. Orlov, G.L. Snider, P.J. Fay, Operation of single-walled carbon nanotube as a radio-frequency single-electron transistor. Nanotechnology **18**, 445203 (2007)
10. J.S. Bunch, A.M. Zande, S.S. Verbridge, I.W. Frank, D.M. Tanenbaum, J.M. Parpia, H.G. Craighead, P.L. McEuen, Electromechanical resonators from graphene sheets. Science **315**, 490 (2007)
11. K. Nishiguchi, A. Fujiwara, Y. Ono, H. Inokawa, Y. Takahashi, Room-temperature-operating data processing circuit based on single-electron transfer and detection with metal-oxide-semiconductor field-effect transistor technology. Appl. Phys. Lett. **88**, 183101 (2006)
12. J. Salfi, I.G. Savelyev, M. Blumin, S.V. Nair, H.E. Ruda, Direct observation of single-charge-detection capability of nanowire field-effect transistors. Nat. Nanotechnol. **5**, 737 (2010)
13. D. Reilly, C. Marcus, Fast single-charge sensing with a rf quantum point contact. Appl. Phys. Lett. **91**, 162101 (2007)
14. G. Podd, S. Angus, D. Williams, A. Ferguson, Charge sensing in intrinsic silicon quantum dots. Appl. Phys. Lett. **96**, 082104 (2010)
15. S. Angus, A. Ferguson, S. Dzurak, G. Clark, A silicon radio-frequency single electron transistor. Appl. Phys. Lett. **92**, 112103 (2008)
16. I. Ahmed, J.A. Haigh, S. Schaal, S. Barraud, Y. Zhu, Ch. Lee, M. Amado, J.W.A. Robinson, A. Rossi, J.J.L. Morton, M.F. Gonzalez-Zalba, Radio-frequency capacitive gate-based sensing. Phys. Rev. Appl. **10**, 014018 (2018)
17. A.N. Cleland, M.L. Roukes, A nanometre-scale mechanical electrometer. Nature **392**, 160 (1998)
18. P. Häkkinen, A. Isacsson, A. Savin, J. Sulkko, P. Hakonen, Charge sensitivity enhancement via mechanical oscillation in suspended carbon nanotube devices. Nano Lett. **15**, 1667–1672 (2015)
19. J. Lee, Y. Zhu, A. Seshia, Room temperature electrometry with SUB-10 electron charge resolution. J. Micromech. Microeng. **18**, 025033 (2008)

20. Y. Zhu, J. Lee, A. Seshia, A resonant micromachined electrostatic charge sensor. IEEE Sens. J. **8**, 1499 (2008)
21. G. Jaramillo, C. Buffa, M. Li, F.J. Brechtel, G. Langfelder, D.A. Horsley, MEMS electrometer with femtoampere resolution for aerosol particulate measurements. IEEE Sens. J. **13**, 2993 (2013)
22. J. Jalil, Y. Ruan, Y. Zhu, Room-temperature sensing of single electrons using vibrating-reed electrometer in silicon-on-glass technology. IEEE Electron Device Lett. **39**, 1928 (2018)
23. P. Thiruvenkatanathan, J. Yan, A.A. Seshia, Ultrasensitive mode-localized micromechanical electrometer, in *Proceedings IEEE International Frequency Control Symposium* (2010), pp. 91–96
24. H. Zhang, J. Huang, W. Yuan, H. Chang, A high-sensitivity micromechanical electrometer based on mode localization of two degree-of-freedom weakly coupled resonators. J. Microelectromech. Syst. **25**, 937 (2016)
25. J. Yang, H. Kang, H. Chang, A micro resonant electrometer with 9-electron charge resolution in room temperature, in *Proceedings IEEE Micro Electro Mechanical Systems (MEMS)* (2018), pp. 67–70
26. J. Lee, B. Bahreyni, A.A. Seshia, An axial strain modulated double-ended tuning fork electrometer. Sens. Actuator A Phys. **148**, 395–400 (2008)
27. D. Chen, J. Zhao, Y. Wang, J. Xie, An electrostatic charge sensor based on micro resonator with sensing scheme of effective stiffness perturbation. J. Micromech. Microeng. **27**, 065002 (2017)
28. D. Chen, J. Zhao, Y. Wang et al., Electrostatic charge sensor based on micro resonator with sensing scheme of effective stiffness perturbation, in *2017 IEEE 30th International Conference on Micro Electro Mechanical Systems (MEMS)* (IEEE, 2017), pp. 1208–1211
29. D. Chen, X. Liu, Y. Wang et al., High sensitivity micro electrometer based on clamped free curved beams resonator with weakened nonlinearity, in *2018 IEEE Micro Electro Mechanical Systems (MEMS)* (IEEE, 2018), pp. 1092–1095
30. M. Sansa, E. Sage, E. Bullard, M. Gély, T. Alava, E. Colinet, A. Naik, L. Villanueva, L. Duraffourg, M. Roukes, G. Jourdan, S. Hentz, Frequency fluctuations in silicon nano-resonators. Nat. Nanotechnol. **11**, 552 (2016)
31. Y.S. Patil, S. Chakram, L. Chang et al., Thermomechanical two-mode squeezing in an ultrahigh-Q membrane resonator. Phys. Rev. Lett. **115**(1), 017202 (2015)
32. S.S. Verbridge, H.G. Craighead, J.M. Parpia, A megahertz nanomechanical resonator with room temperature quality factor over a million. Appl. Phys. Lett. **92**(1), 013112 (2008)
33. G. Sobreviela, X. Zou, C. Zhao et al., An ultra-high-quality factor silicon disk resonator, in *2019 20th International Conference on Solid-State Sensors, Actuators and Microsystems & Eurosensors XXXIII (TRANSDUCERS & EUROSENSORS XXXIII)* (IEEE, 2019), pp. 527–530
34. S.J. Wong, C.H.J. Fox, S. McWilliam, Thermoelastic damping of the in-plane vibration of thin silicon rings. J. Sound Vib. **293**(1–2), 266–285 (2006)
35. A.N. Cleland, M.L. Roukes, A nanometre-scale mechanical electrometer. Nature **392**(6672), 160 (1998)
36. A.K. Naik, M.S. Hanay, W.K. Hiebert et al., Towards single-molecule nanomechanical mass spectrometry. Nat. Nanotechnol. **4**(7), 445 (2009)
37. Y. Wang, Y. Wang, W. Liu et al., An aerosol sensor for PM1 concentration detection based on 3D printed virtual impactor and SAW sensor. Sens. Actuator A Phys. **288**, 67–74 (2019)
38. H. Ding, J. Zhao, B.F. Ju et al., A high-sensitivity biaxial resonant accelerometer with two-stage microleverage mechanisms. J. Micromech. Microeng. **26**(1), 015011 (2015)
39. W. Zhang, J.E.Y. Lee, Frequency-based magnetic field sensing using Lorentz force axial strain modulation in a double-ended tuning fork. Sens. Actuator A Phys. **211**, 145–152 (2014)
40. D. Chen, Y. Wang, X. Chen et al., Temperature-frequency drift suppression via electrostatic stiffness softening in MEMS resonator with weakened duffing nonlinearity. Appl. Phys. Lett. **114**(2), 023502 (2019)
41. X. Liu, D. Chen, D. Yang et al., A computational piezoelectric micro-machined ultrasonic transducer toward acoustic communication. IEEE Electron Device Lett. **40**(6), 965–968 (2019)

42. A. Rahafrooz, S. Pourkamali, High-frequency thermally actuated electromechanical resonators with piezoresistive readout. IEEE Trans. Electron Devices **58**(4), 1205–1214 (2011)
43. J.E.-Y. Lee, A. Seshia, Parasitic feedthrough cancellation techniques for enhanced electrical characterization of electrostatic microresonators. Sens. Actuator A Phys. **156**, 36 (2009)
44. Y. Xu, J.E.-Y. Lee, Single-device and on-chip feedthrough cancellation for hybrid MEMS resonators. IEEE Trans. Ind. Electron. **59**, 4930 (2012)
45. H. Zhang, W. Yuan, Y. Hao, H. Chang, Influences of the feedthrough capacitance on the frequency synchronization of the weakly coupled resonators. IEEE Sens. J. **11**, 6081 (2015)
46. V. Kaajakari, T. Mattila, A. Oja et al., Nonlinear limits for single-crystal silicon microresonators. J. Microelectromech. Syst. **13**(5), 715–724 (2004)
47. I. Kozinsky, H.W.C. Postma, I. Bargatin et al., Tuning nonlinearity, dynamic range, and frequency of nanomechanical resonators. Appl. Phys. Lett. **88**(25), 253101 (2006)
48. D. Chen, Y. Wang, Y. Guan et al., Methods for nonlinearities reduction in micromechanical beams resonators. J. Microelectromech. Syst. **27**(5), 764–773 (2018)
49. T.A.W. Roessig, Integrated MEMS tuning fork oscillators for sensor applications (1999)
50. C. Zuo, N. Sinha, J. Van der Spiegel et al., Multifrequency pierce oscillators based on piezoelectric AlN contour-mode MEMS technology. J. Microelectromech. Syst. **19**(3), 570–580 (2010)
51. L. He, Y.P. Xu, M. Palaniapan, A state-space phase-noise model for nonlinear MEMS oscillators employing automatic amplitude control. IEEE Trans. Circuits Syst. I Regul. Pap. **57**(1), 189–199 (2009)
52. C. Zhao, G. Sobreviela, M. Pandit et al., Experimental observation of noise reduction in weakly coupled nonlinear MEMS resonators. J. Microelectromech. Syst. **26**(6), 1196–1203 (2017)
53. G. Sobreviela, C. Zhao, M. Pandit et al., Parametric noise reduction in a high-order nonlinear MEMS resonator utilizing its bifurcation points. J. Microelectromech. Syst. **26**(6), 1189–1195 (2017)
54. M. Pandit, C. Zhao, G. Sobreviela et al., Coupled nonlinear MEMS resonators for sensing, in *2018 IEEE International Frequency Control Symposium (IFCS)* (IEEE, 2018), pp. 1–4
55. A. Cowen, G. Hames, D. Monk, S. Wilcenski, B. Hardy, *SOIMUMPs Design Handbook* (MEMSCAP Inc, Durham, NC, 2011)
56. C.S. Li, L.J. Hou, S.S. Li, Advanced CMOS–MEMS resonator platform. IEEE Electron Device Lett. **33**(2), 272–274 (2011)
57. D. Chen, J. Zhao, Y. Wang, Zh Xu, J. Xie, Sensitivity manipulation on micro-machined resonant electrometer toward high resolution and large dynamic range. Appl. Phys. Lett. **112**, 013502 (2018)
58. D. Chen, J. Zhao, Zh Xu, J. Xie, A micro resonant charge sensor with enhanced sensitivity based on differential sensing scheme and leverage mechanisms. AIP Adv. **6**, 105106 (2016)
59. X.P.S. Su, H.S. Yang, Two-stage compliant microleverage mechanism optimization in a resonant accelerometer. Struct. Multidiscip. Optim. **22**(4), 328–334 (2001)
60. J. Zhao, H. Ding, J. Xie, Electrostatic charge sensor based on a micromachined resonator with dual micro-levers. Appl. Phys. Lett. **106**(23), 233505 (2015)
61. X. Zou, *High Resolution Resonant Accelerometer Based on MEMS Technology* (University of Cambridge, 2014)
62. Y. Wang, H. Ding, X. Le et al., A MEMS piezoelectric in-plane resonant accelerometer based on aluminum nitride with two-stage microleverage mechanism. Sens. Actuator A Phys. **254**, 126–133 (2017)
63. J.C. McCormac, R.E. Elling, *Structural Analysis* (Harper and Row, New York, 1984)
64. M. Manav, G. Reynen, M. Sharma et al., Ultrasensitive resonant MEMS transducers with tuneable coupling. J. Micromech. Microeng. **24**(5), 055005 (2014)
65. D. Chen, Y. Wang, X. Chen et al., Duffing nonlinearity localization via extension energy confinement in an elastic mode semicircular beams resonator. IEEE Electron Device Lett. **40**(2), 314–317 (2018)
66. D. Chen, H. Zhang, J. Sun, M. Pandit et al., A micro resonant electrometer with single-electron charge resolution at room temperature, in Proceedings IEEE Micro Electro Mechanical Systems (MEMS) (2020), pp.182–185

67. D. Chen, H. Zhang, J. Sun, M. Pandit et al., Phase-controlled oscillation in a capacitive nonlinear ring resonator with on-chip feedthrough de-embedding, in Proceedings IEEE Micro Electro Mechanical Systems (MEMS) (2020), pp.773–776
68. L.G. Villanueva, E. Kenig, R.B. Karabalin et al., Surpassing fundamental limits of oscillators using nonlinear resonators. Phys. Rev. Lett. **110**(17), 177208 (2013)
69. D. Antonio, D.H. Zanette, D. López, Frequency stabilization in nonlinear micromechanical oscillators. Nat. Commun. **3**, 806 (2012)
70. D.K. Agrawal, J. Woodhouse, A.A. Seshia, Observation of locked phase dynamics and enhanced frequency stability in synchronized micromechanical oscillators. Phys. Rev. Lett. **111**(8), 084101 (2013)

$$[m_2s^2 + c_2s + (k_2 + k_c)]X_2(s) - k_cX_1(s) = F_2(s) \tag{2b}$$

Applying Cramer's rule, we can obtain:

$$X_1(s) = \frac{F_1(s)H_2(s) + k_cF_2(s)}{H_1(s)H_2(s) - k_c^2} \tag{3a}$$

$$X_2(s) = \frac{F_2(s)H_1(s) + k_cF_1(s)}{H_1(s)H_2(s) - k_c^2} \tag{3b}$$

Where $H_1(s)$ and $H_2(s)$ are given by:

$$H_1(s) = m_1s^2 + c_1s + (k_1 + k_c) \tag{4a}$$

$$H_2(s) = m_2s^2 + c_2s + (k_2 + k_c) \tag{4b}$$

If the resonators are actuated by only F_1 (i.e. $F_2 = 0$), Eqs. (3a) and (3b) become:

$$X_1(s) = \frac{F_1(s)H_2(s)}{H_1(s)H_2(s) - k_c^2} \tag{5a}$$

$$X_2(s) = \frac{k_cF_1(s)}{H_1(s)H_2(s) - k_c^2} \tag{5b}$$

It is not difficult to find that:

$$\frac{X_2(s)}{X_1(s)} = \frac{k_c}{H_2(s)} \tag{6}$$

Now we consider the following two cases.

When the two resonators are identical. In this case, $k_1 = k_2 = k$, $m_1 = m_2 = m$, and $c_1 = c_2 = c$. The angular frequencies of the two fundamental resonant modes can be calculated as:

$$\omega_1 = \sqrt{\frac{k + 2k_c}{m}} \tag{7a}$$

$$\omega_2 = \sqrt{\frac{k}{m}} \tag{7b}$$

Substituting Eqs. (7a) and (7b) into Eq. (6), neglecting damping terms, the amplitude ratio at two fundamental resonant modes are:

$$\frac{X_2(\omega_1)}{X_1(\omega_1)} = -1 \tag{8a}$$

$$\frac{X_2(\omega_2)}{X_1(\omega_2)} = 1 \tag{8b}$$

When there is a slight mismatch in the stiffness of the two resonators. In this case, we assume $k_1 = k$ and $k_2 = k(1 + \delta k)$, and δk is the mismatch coefficient. The other parameters are still identical. Note that this case is essentially what happens when electron charges are introduced, altering the stiffness of a resonator, but not other parameters. Substituting k_1 and k_2 into Eqs. (5a) and (5b), the angular frequencies of the two fundamental modes in this case can be derived and expressed as:

$$\omega_1 = \sqrt{\frac{2(k + k_c) + k\delta k - \sqrt{(k\delta k)^2 + 4k_c^2}}{2m}} \tag{9a}$$

$$\omega_2 = \sqrt{\frac{2(k + k_c) + k\delta k + \sqrt{(k\delta k)^2 + 4k_c^2}}{2m}} \tag{9b}$$

With these two frequencies, and neglecting the damping terms, the amplitude ratios at these two frequencies, $\chi(\omega_1)$ and $\chi(\omega_2)$, can be calculated as:

$$\chi(\omega_1) = \frac{X_2(\omega_1)}{X_1(\omega_1)} = -\frac{k\delta k}{2k_c} - \sqrt{\left(\frac{k\delta k}{2k_c}\right)^2 + 1} \tag{10a}$$

$$\chi(\omega_2) = \frac{X_2(\omega_2)}{X_1(\omega_2)} = -\frac{k\delta k}{2k_c} + \sqrt{\left(\frac{k\delta k}{2k_c}\right)^2 + 1} \tag{10b}$$

It can be clearly observed from Eqs. (10a) and (10b), as well as Fig. 2, that the amplitude ratio is dependent on the stiffness mismatch between the two resonators. Taking the derivative of the amplitude ratio with respect to the stiffness mismatch, the scale factor of the amplitude ratio to stiffness mismatch, $\mathcal{C}_\chi$, can be expressed as:

$$\mathcal{C}_\chi(\omega_1) = \frac{d[\chi(\omega_1)]}{d(\delta k)} = -\frac{k}{2k_c} - \frac{k^2\delta k}{4k_c^2}\left[\left(\frac{k\delta k}{2k_c}\right)^2 + 1\right]^{-\frac{1}{2}} \tag{11a}$$

$$\mathcal{C}_\chi(\omega_2) = \frac{d[\chi(\omega_2)]}{d(\delta k)} = -\frac{k}{2k_c} + \frac{k^2\delta k}{4k_c^2}\left[\left(\frac{k\delta k}{2k_c}\right)^2 + 1\right]^{-\frac{1}{2}} \tag{11b}$$

From the above expressions, it can be observed that the scale factor is a nonlinear function of the mismatch. However, a careful examination of Eqs. (11a) and (11b) reveals that the scale factor can be approximated with three linear functions in three regions:

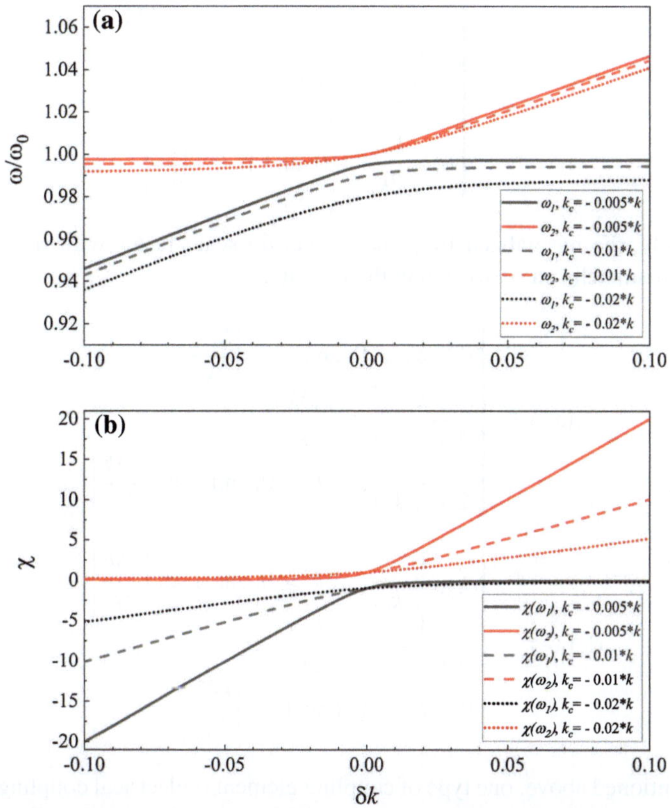

Fig. 2 Theoretically calculated responses with respect to a stiffness mismatch, δk, for electrically coupled devices (i.e. $k_c < 0$): **a** normalized resonant frequency variations within a 2DoF mode-localized sensor, ω/ω_0; **b** amplitude ratio changes within a 2DoF mode-localized sensor, X_2/X_1. It can be observed that the amplitude ratio changes more drastically than that in the resonant frequencies. Additionally, it can be seen that the changes in amplitude ratio is a function of coupling strength, the smaller the k_c, the more drastic change in the amplitude ratio

$$\mathcal{C}_\chi(\omega_1) \approx \begin{cases} 0, & \frac{k\delta k}{2k_c} < 0, \text{ and } 1 \ll \left|\frac{k\delta k}{2k_c}\right| \\ -\frac{k}{2k_c}, & \left|\frac{k\delta k}{2k_c}\right| \ll 1 \\ -\frac{k}{k_c}, & \frac{k\delta k}{2k_c} > 0, \text{ and } 1 \ll \left|\frac{k\delta k}{2k_c}\right| \end{cases} \tag{12a}$$

$$\mathcal{C}_{\chi}(\omega_2) \approx \begin{cases} -\dfrac{k}{k_c}, & \dfrac{k\delta k}{2k_c} < 0, \text{ and } 1 \ll \left|\dfrac{k\delta k}{2k_c}\right| \\ -\dfrac{k}{2k_c}, & \left|\dfrac{k\delta k}{2k_c}\right| \ll 1 \\ 0, & \dfrac{k\delta k}{2k_c} > 0, \text{ and } 1 \ll \left|\dfrac{k\delta k}{2k_c}\right| \end{cases} \tag{12b}$$

Similarly, the normalized frequency's (i.e. $\hat{\omega}$) scale factor with respect to the stiffness mismatch can be written in three regions[2]:

$$\mathcal{C}_{\hat{\omega}}(\omega_1) \approx \begin{cases} 0, & \delta k > 0, \text{ and } 1 \ll \left|\dfrac{k\delta k}{2k_c}\right| \\ \dfrac{k}{4(k+k_c)}, & \left|\dfrac{k\delta k}{2k_c}\right| \ll 1 \\ \dfrac{k}{2(k+k_c)}, & \delta k < 0, \text{ and } 1 \ll \left|\dfrac{k\delta k}{2k_c}\right| \end{cases} \tag{13a}$$

$$\mathcal{C}_{\hat{\omega}}(\omega_2) \approx \begin{cases} \dfrac{k}{2(k+k_c)}, & \delta k > 0, \text{ and } 1 \ll \left|\dfrac{k\delta k}{2k_c}\right| \\ \dfrac{k}{4(k+k_c)}, & \left|\dfrac{k\delta k}{2k_c}\right| \ll 1 \\ 0, & \delta k < 0, \text{ and } 1 \ll \left|\dfrac{k\delta k}{2k_c}\right| \end{cases} \tag{13b}$$

As mentioned above, one type of coupling element is electrical coupling. It offers the advantage of high tunability, hence it is widely used in mode-localized sensors. Here, if electrical coupling is used, meaning that $k_c < 0$, the scale factor of amplitude ratio and normalized frequency can be directly compared side-by-side. From Eqs. (12a) to (13b), the improvement in scale factor can be approximated as:

$$\frac{\mathcal{C}_{\chi}(\omega_1)}{\mathcal{C}_{\hat{\omega}}(\omega_2)} \approx -\frac{2(k+k_c)}{k_c} \tag{14}$$

From this equation, it can be seen that the larger ratio of k/k_c, the higher scale factor improvement. reason why *weakly coupled resonators* are used within a mode-localized sensor—to enhance the sensitivity. In Fig. 3, the scale factor improvement utilizing the *mode-localization* effect can be clearly observed through a simulation. Generally, the ratio of k/k_c is in the order of a few hundred to a few thousand, meaning that the improvement in scale factor is typically two to three orders of magnitude.

[2]Since the amplitude ratio is a dimensionless value, to fairly compare the two scale factors, it is necessary to normalize the scale factor [20].

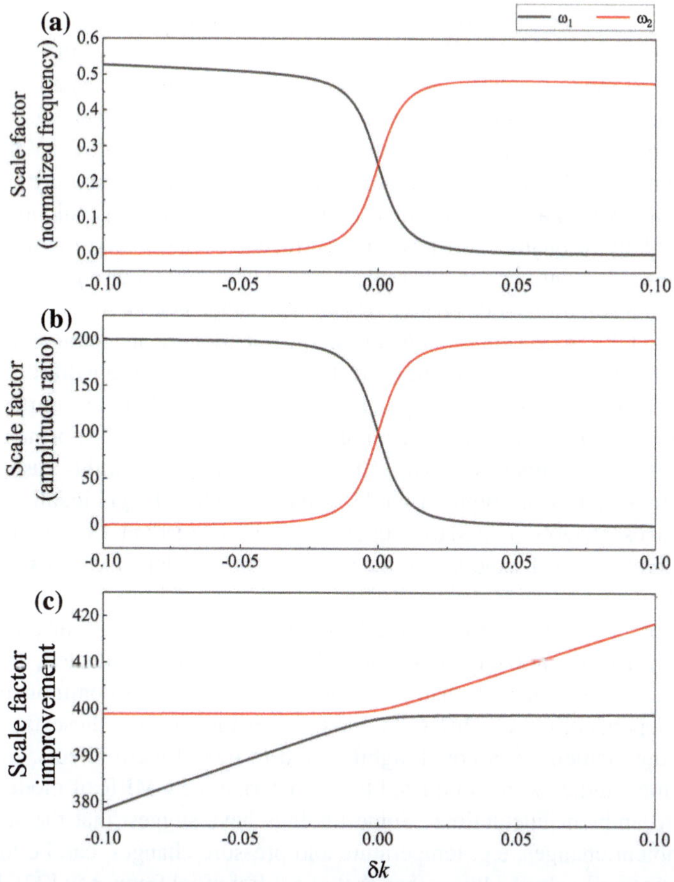

Fig. 3 Theoretically calculated scale factors with respect to a stiffness mismatch, δk, for electrically coupled devices, $k_c = -0.005 \times k$: **a** normalized resonant frequency's scale factor; **b** amplitude ratio's scale factor and **c** the scale factor is improved by approximately 399 times [which can also be estimated using Eq. (14)] utilizing the *mode-localization* effect

2.2 Common Mode Rejection Capabilities

For virtually all sensors, the common mode rejection capabilities is one of the key specifications, as it indicates how the sensor performs when common mode signals, most typically the ambient fluctuations, are present. This concept is similar to the common mode rejection ratio (CMRR) in integrated operational amplifiers, hence we "borrow" the definition:

Definition 1

$$\text{CMRR} = \frac{\text{Scale factor to differential input}}{\text{Scale factor to common mode input}} \tag{15}$$

To fairly compare the two scale factors, it is necessary to convert both the differential and common mode input to a change in the stiffness of the resonator. In the context of mode-localized electrometers, an electrostatic force generated by the charge injection results in a change in one resonator's stiffness; this is considered the differential input. The scale factor has been discussed in Sect. 2.1. On the other hand, common mode charge injection, ambient fluctuations, temperature fluctuations in particular, as well as identical stress fatigue in the resonators, could also result in changes of both resonators' stiffness; these effects are considered the common mode input. In the following section, the scale factor to common mode signal is discussed.

Consider a common mode change results in a change in both resonators' stiffness, $k' = k\Delta k$. This can be a common mode charge injection, or an identical drift in both resonators' stiffness. Following the derivations in Sect. 2.1, it is not difficult to obtain that for a given stiffness change input, i.e. a constant $k\delta k$,[3] both the amplitude ratio, χ is a function of k_c, while independent of Δk. On the contrary, the normalized mode frequencies, ω/ω_0, depends on Δk. This is shown in Fig. 4. In this case, the scale factor to these types of common mode signals is reduced significantly for mode-localized sensors, as compared to frequency shift based readout. Combining with the high differential signal scale factor, the CMRR for mode-localized sensors is much improved.

However, in practice, the more usual common mode signals are the ambient fluctuations. In this case, not only the stiffness of the resonators are affected, the coupling element, k_c, is also subject to changes. As mentioned above, the common mode scale factor is dependent on k_c. Due to the complex nature of how these factors affect the coupling element, it is not straightforward to model these effects. In the past, experimental studies were conducted to characterize the CMRR of mode-localized sensors to ambient fluctuations. Some findings have suggest that the scale factor to the ambient changes, e.g. temperature and pressure changes, can be lower [21], but not necessarily always lower than that of a differential frequency [22]. However, thanks to the high differential scale factor, the resulting CMRR is typically improved by approximately two orders of magnitude nonetheless.

2.3 *Operation of mDoF Mode-Localized Sensors*

Along the train of thoughts that by adding one more resonator to the conventional single resonator, the scale factor and CMRR can be improved significantly, one curious mind might be wondering: "*What would happen if more than two resonators are weakly coupled together?*" In this section, we briefly discuss the potential possibility of using a mDoF mode-localized sensors, which consist of m resonators. In the literature, there already exists 3DoF mode-localized sensors, as shown in Fig. 5.

[3] This is generally true for a force input, as the stiffness change induced by the axial stress is generally much larger than that caused by strain variations.

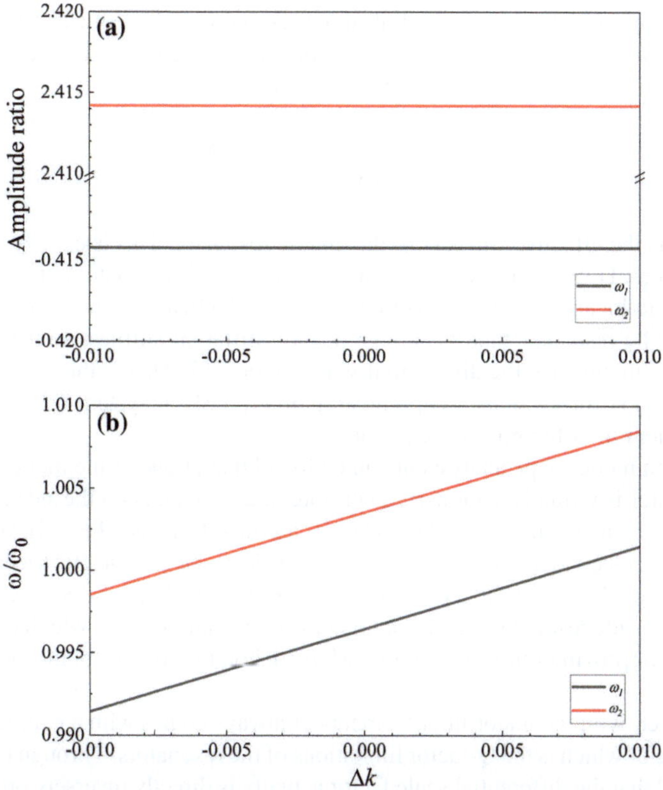

Fig. 4 Theoretically calculated amplitude ratio, χ and normalized mode frequencies, ω/ω_0, as a function of common mode stiffness change, Δk. This is calculated under the condition of a weakly coupling, $k_c = -0.005k$, and a constant differential input $k\delta k$

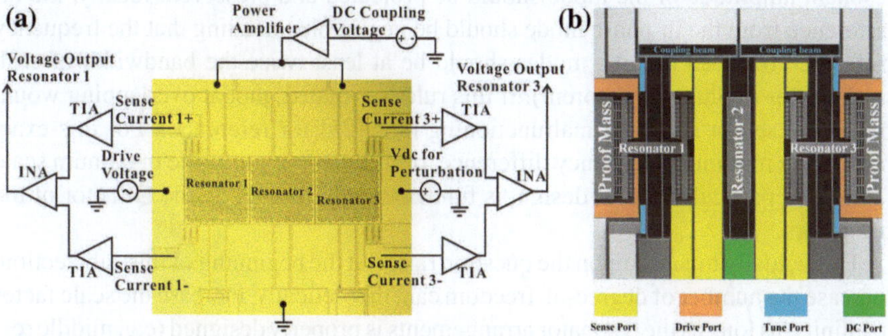

Fig. 5 **a** A typical electrically coupled 3DoF mode-localized sensor consisting of two identical resonators on the side, and a resonator with higher stiffness in the middle, which are coupled by electrostatic forces [23]. **b** A mechanically coupled 3DoF mode-localized sensor consisting of two identical resonators on each side, and a slightly different resonator in the middle [24]

Assuming the effective mass of all three resonators are identical, one can theoretically estimate the differential scale factor improvement as compared to a 2DoF mode-localized sensor [23], which is expressed by:

$$\frac{\mathcal{C}_{\chi}(\text{3DoF})}{\mathcal{C}_{\chi}(\text{2DoF})} = \frac{k_2 - k + k_c}{k_c} \tag{16}$$

where k_2 is the effective stiffness of the middle resonator. This is exactly the reason why, unlike 2DoF mode-localized sensors, where both resonators are identical, a 3DoF mode-localized sensor generally has two identical resonators on each side and a middle resonator that has a significantly different stiffness to the other two resonators: to enhance the differential scale factor [25]. That is the reason why for the identical stiffness, k and coupling strength k_c, a 3DoF system has a significant improvement in differential scale factor.

Through mode shape analysis, it can be found that at one of the modes, the middle resonator is virtually stationary. Therefore, the resonators on the side essentially vibrate between two stationary electrodes. If carefully designed, the 2nd order nonlinearity of the electrostatic force within the drive electrode and the electrical coupling can be canceled out. Hence, another advantage of the 3DoF mode-localized sensor is that the two side resonators can have higher resonant amplitudes, which can be beneficial for improving sensor resolution. More elaboration on this point can be found in [23].

However, with the major improvements, it always comes with a couple of drawbacks, one of which is the Q-factor limitations of the resonators. Through research, it was found that the differential scale factor actually is directly inversely proportional to the frequency difference between an in-phase mode and an anti-phase mode. It should be pointed out that if the desired mode of operation is typically the anti-phase mode; therefore, to ensure proper operation of the 3DoF mode-localized sensor, the resonant amplitude of the mode should be protected and preserved. Ideally, the interference from the in-phase mode should be negligible, meaning that the frequency difference between the two modes should be at least twice the bandwidth of each mode (refer to Shannon theorem). If this rule is violated, modal overlapping would occur, the sensor could be malfunctioning (see [25] for reference). For this exact reason, the minimum frequency difference, or more importantly, the maximum scale factor of a particular sensor design, is fundamentally limited by the Q-factor of the resonators.

This actually touched upon the question raised at the beginning of this sub-section. Increase the number of degree-of-freedom can, theoretically, increase the scale factor to infinity, as long as the resonator arrangements is properly designed (e.g. middle resonator has a higher stiffness than the other two resonators in a 3DoF mode-localized sensor). However, this statements is only valid in theory and ideal cases where the Q-factor of the resonators are infinite. In practice, all resonators have energy loss mechanisms, meaning that an infinite Q-factor is effectively impossible. In a 3DoF mode-localized sensor, an invisible ceiling of Q-factor is already reached, meaning that any higher number of DoF ($m \geq 4$) is not practical in the sense of improving differential scale factor. However, novel designs reported in recent publications have shown that 4DoF system can be useful in improving the scale factor of an accelerometer [26, 27]

But, the high number of degree-of-freedom can be extremely helpful in identifying the location of perturbation [28]. Therefore, if the required specs include perturbation location identification, it is still useful to design a *m*DoF mode-localized sensor.

For the following sections, unless otherwise specified, the default system for discussion is a 2DoF resonant system.

3 Design Considerations for Mode-Localized Electrometers

In this section, we shall focus on the practical implementations of MEMS mode-localized electrometers. A typical MEMS mode-localized electrometer [29] should consist of a MEMS coupled resonator structure adopted for the electrometer application, the associated electronic circuitry utilized for monitoring the resonators' motional current, as well as locking the resonator to its resonant frequencies. In addition, there are a few other practical considerations in implementing a mode-localized electrometer, such as the resolution enhancement. Hence, in the following sections, we shall discuss the practical implementation in three different aspects.

3.1 Device Designs of Mode-Localized Electrometers

One viable device implementation of a 2DoF MEMS mode-localized electrometer is shown in Fig. 6b. The device consists of two identical resonators of the double-ended-tuning-fork (DETF) type. The coupling in this case is of mechanical coupling type.

3.1.1 Resonator Designs

The choice of resonator type here is not arbitrary. Firstly, as discussed above, the Q-factor of the resonators ultimately limits the maximum scale factor of the mode-localized sensors. Therefore, to maximize the performance of the mode-localized electrometer, the resonators within the system should have high Q-factor. Thanks to the symmetrical design, the anchor loss of the DETF resonators are generally extremely low at its anti-phase mode. Hence, the Q-factor of the resonators are among the highest in the resonator designs. Secondly, the choice of the resonator type should facilitate the application of the charge injection. On this point, the electrostatic force resulted from a charge input can easily modulate the axial stress of the DETF type of resonator. Combining these two points, the DETF type of resonator is ideal for the practical implementation of mode-localized electrometers.

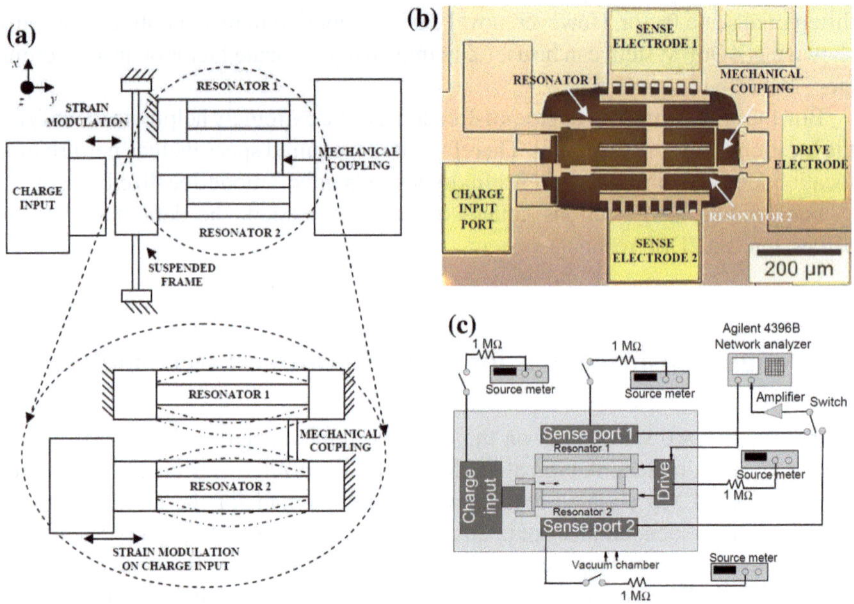

Fig. 6 A typical implementation of a 2DoF mode-localized electrometer [29]: **a** the schematic of a mechanically coupled 2DoF mode-localized electrometer. It also shows the operation principle of the mode-localized electrometer, through the stress modulation caused by charge injection; **b** an optical micrograph of the device, showing an example of practical construction of the electrometer; **c** the associated open-loop test configuration of the electrometer

To further improve the quality factor, one might think of using other types of high Q-factor resonators, including bulk mode resonators. However, it is generally difficult to modulate the stiffness of these types of resonators, making them less sensitive to charge injections. Therefore, the bulk mode resonators are generally not recommended for mode-localized electrometers implementations.

However, there are still many other types of resonators that could be utilized for the application here, such as ring resonators [30] and DETF type resonators with slots to minimize thermoelastic damping [31].

It should be pointed out that the improvement of Q-factor is beneficial for achieving high scale factor for mode-localized electrometers, as well as for reducing the phase noise within the close-loop configuration, which will be discussed in Sect. 3.2. At the same time, increasing Q-factor can increase the effect of nonlinearity in the resonators, hence reducing the critical vibration amplitude [32]. As shall be discussed in Sect. 3.4, lower vibration amplitude can result in lower sensor resolution. For this reason, there is a trade-off in choosing the resonator topology as well as designing the dimensions of the resonators; and a balance between scale factor and resolution should be achieved.

Additionally, for these flexural resonators, a DC potential difference is applied between the body of the resonators and the drive/sense electrodes, polarizing the

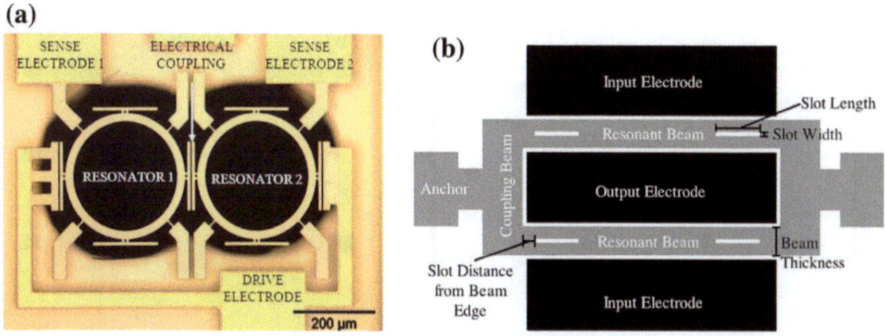

Fig. 7 Different types of resonators that could potentially be used for mode-localized electrometers: **a** a ring resonator, whose flexural wine glass mode could be utilized [30]; **b** a DETF resonator with slots, whose flexural modes is utilized [31]

resonators. On top of that, a drive electrode is used to transduce an AC electrical signal into an electrostatic force between the resonators and the drive electrode, thus actuating the resonant system. When the frequency of the AC electrical signal is close to the desired operational frequency, i.e. anti-phase mode in this case (see Fig. 6a), the vibration amplitudes reach their maximum. A third electrode as the output electrode is necessary to transduce the motion of the resonators into motional current, then to be picked up by the front-end electronics. The method to drive the resonators at their resonant frequency, as well as front-end electronics, shall be discussed in Sect. 3.4 (Fig.7).

Furthermore, it should be pointed out that typically, the mode-localized sensors are placed in a vacuum environment to achieve the highest Q-factor possible; the vacuum environment can be a vacuum chamber [33], a device level vacuum packaging [24] or a wafer level vacuum package [34].

3.1.2 Charge Injection Approach

For MEMS resonant electrometers, charge injection is introduced to cause a shift in the electrostatic force between the injection electrode and the resonator. This electrostatic force then modulates the stiffness of the resonator. However, there are two approaches to apply this stiffness change: (1) the electrostatic force is applied axially, meaning that the electrostatic force is perpendicular to the direction of resonator's vibration; (2) the electrostatic force is applied in the direction parallel to the direction of resonator's vibration. The schematics of the two approaches are shown in Fig. 8.

For the first approach (see Fig. 8a), the electrostatic force is applied axially, perpendicular to the direction of vibration. With this approach, the resulting stiffness change is through modulating the axial stress of the resonators, and the effect is given by Roessig [37]:

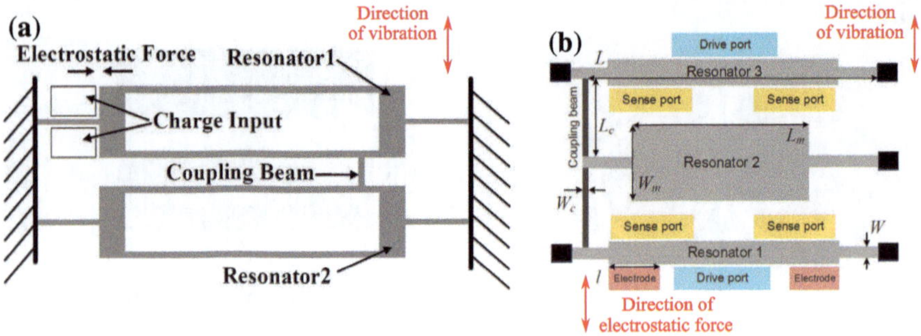

Fig. 8 Different charge injection approaches can be used for mode-localized electrometers: **a** axially applied electrostatic force [35]; **b** the electrostatic force applied along the direction of vibration [36]. It should be pointed out that the flexural modes of the resonators are all in-plane

$$\delta k = \frac{T \int_0^1 \left(\frac{\partial u(x)}{\partial x} \right)^2 \mathrm{d}x}{kL} \tag{17}$$

where k, L and $u(x)$ is the stiffness of the resonator, length of the resonator and the mode shape of the resonator along the axis of the resonator, respectively. T is the electrostatic force given by:

$$T = \frac{\varepsilon_0 A}{2d^2} \left(V_0 - \frac{qd}{\varepsilon_0 A} \right)^2 \approx \frac{qV_0}{d} \tag{18}$$

where ε_0, A and d is the dielectric constant of vacuum, the overlapping area and the gap of the capacitor between the charge injection electrode and the resonator. V_0 is the DC polarization voltage on the body of the resonator.[4] The relationship between the scale factor and the parameters can be derived:

$$\frac{\mathrm{d}\chi}{\mathrm{d}q} \propto \frac{V_0}{d} \frac{\int_0^1 \left(\frac{\partial u(x)}{\partial x} \right)^2 \mathrm{d}x}{kL} \tag{19}$$

It can be seen that to increase the scale factor, the polarization voltage should be improved, while the gap between the charge injection electrode and the resonator should be reduced.

For the second approach (see Fig. 8b), the electrostatic force is dependent on the resonator displacement; thus, it can be perceived as an electrostatic spring, whose spring constant is dependent on the charge injection. The normalized spring constant change due to the charge injection can be estimated as:

[4]The approximation is valid when $\frac{qd}{\varepsilon_0 A} < V_0$, which is true for small charge injections.

$$\delta k = \frac{\varepsilon_0 A}{kd^3}\left(V_0 - \frac{qd}{\varepsilon_0 A}\right)^2 \approx \frac{2qV_0}{kd^2} \tag{20}$$

Hence, the relationship between the scale factor and dimensions using this approach can be expressed as:

$$\frac{\mathrm{d}\chi}{\mathrm{d}q} \propto \frac{V_0}{kd^2} \tag{21}$$

Typically, the length of the resonator, L is in the order of a few hundred microns (μm), the mode shape dependent constant $\int_0^1 \left(\frac{\partial u(x)}{\partial x}\right)^2 \mathrm{d}x$ is approximately 5 for fixed-fixed beams. On a typical design, the gap between the charge injection electrode and the resonator, d, is typically a few microns (μm) or even smaller. By comparing Eqs. (19) and (21), assuming identical k and V_0, it can be clearly concluded that the scale factor of second approach should be higher than that of the first approach.

However, for the second approach, the full range of sensing can be limited, as the effect of dynamic pull-in can occur if large amount of charge is injected to the sense electrode. In this sense, for the large dynamic range, the first approach can be useful.

3.2 Electronics Implementation of Mode-Localized Electrometers

3.2.1 Front-End Amplifier Design

As discussed above, the motional current of the resonators need to be picked up by a front-end transimpedance amplifier (TIA). Since motional current of the resonators are generally extremely weak, in the order of nA (i.e. 10^{-9} A), the front-end TIA should be carefully designed to be able to pick up the current. Furthermore, the front-end TIA is the main contributor of the total electronics noise (as the gain of this stage is typically the highest in the front-end analog circuitry), which ultimately limits the current resolution of the mode-localized electrometers. For this reason, the noise of the front-end TIA should be minimized, to achieve a high signal-to-noise ratio (SNR). A typical low noise TIA topology is shown in Fig. 9 [38].

Since the input signal of the TIA is a current, hence it is fair to estimate the SNR through calculations of the equivalent current noise referred at the input of the "MEMS input" port (see Fig. 9). The input-referred noise current can be estimated as [38]:

$$\begin{aligned}\left[\overline{i_n}(f)\right]^2 &\approx \left[\overline{i_{u1}}(f)\right]^2 + \left[\overline{e_{u1}}(f)\right]^2\left[2\pi f\left(C_f + C_{p,d} + C_{p,u1}\right)\right]^2 \\ &\quad + \left[\overline{e_{u1}}(f)\right]^2\left(\frac{1}{R_b}\right)^2 + \frac{4kT}{R_b}\end{aligned} \tag{22}$$

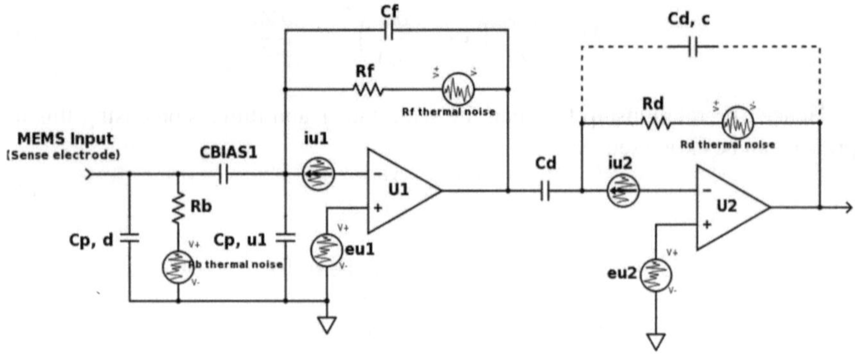

Fig. 9 An typical low-noise front-end amplifier for mode-localized electrometers [38]

where $\overline{i_n}(f)$, $\overline{i_{u1}}(f)$ and $\overline{e_{u1}}(f)$ denotes the frequency (f) dependent noise current referred at the input, the current noise and voltage noise spectral density (NSD) of the first stage operational amplifier, respectively. C_f, $C_{p,d}$ and $C_{p,u1}$ are the capacitance value of the feedback capacitor, parasitic capacitance of the device and the first stage amplifier, respectively. R_b is the bias resistor. All of the above denotations are also shown in Fig. 9.

From Eq. (22), it can be concluded that the input-referred noise current depends critically on the op-amp noise performance. It is therefore imperative to implement the front-end TIA using ultra-low noise op-amps. Furthermore, the input-referred noise current also depends on the input impedance seen from the MEMS input port. Hence, all the parasitic capacitance connecting to the ground, $C_{p,d} + C_{p,u1}$, should be minimized, and the value of the bias resistor, R_b, should be maximized. Additionally, the feedback capacitance, C_f, should be minimized for the benefit of noise performance, as well as for the signal gain. Another side note on the selection of operation amplifiers is that the input bias current of the op-amp should be low. This is because the input-bias current will result in a DC error at the output of the TIA, and its temperature dependence will result in a temperature drift error at the output. To reduce this DC error, it is *NOT recommended* to include a balance RC network at the positive input port of the op-amp, as it will cause significant elevation of the input-referred noise current. Therefore, it is desired to have an op-amp that has low input-bias current. To summarize, the ideal op-amp for this purpose should have **low current and voltage noise**, **low input parasitic capacitance** and **low input-bias current**. For reference, an example op-amp satisfying these requirements can be ADA4817-1 from Analog Devices Inc. Regarding to the associated passive components, it is desired to have high resistance and low capacitance.

As for the second stage, it should be pointed out that the estimation in Eq. (22) is based on the assumption that the gain of the first stage is high (i.e. typically > a few MΩ), and the noise voltage and current of the second stage is low (i.e. $\overline{e_{u2}}(f) < 10\,\text{nV/Hz}^{1/2}$, $\overline{i_{u2}}(f) < 10\,\text{fA/Hz}^{1/2}$). Hence, these are the basic guidelines of selecting the second op-amp. For choosing the associated passive components, special attention should be paid to the closed-loop stability of the second stage op-amp. Meanwhile, a high gain of the second stage and a satisfactory phase of the entire TIA (typically within $180^\circ \pm 5^\circ$) should not be sacrificed. The phase of the TIA should be ideally as close as possible to 180°, so that no further phase shifting circuitry is required at later stages (to reduce error sources).

3.2.2 Closed-Loop Configuration Designs

In Sect. 2.1, the amplitude ratios being discussed are obtained at the resonant frequency of the weakly coupled resonator system. One approach of obtaining this resonant amplitude of the mode-localized electrometers is through an open-loop approach [33], as shown in Fig. 10a. It is easy to implement, without the necessity to design more complicated feedback electronics. However, this approach relies on the frequency sweep to locate the resonant frequency. Therefore, more often than not, the resolution is not high enough; more importantly, this approach is not suitable for real-time measurements. For practical electrometer applications where real-time charge injection information is required, a closed-loop configuration is essential [39]. A typical close-loop test configuration is shown in Fig. 10b.

As can be seen from Fig. 10b, the closed-loop configuration of the mode-localized electrometers is essentially closed-loop setup for one side (resonator on the left), and open-loop read out only for the other side (resonator on the right). The considerations behind this test setup are based on the theoretically calculated and experimentally observed fact that the two resonators resonate at identical frequencies. It is therefore only necessary to make sure one master resonator is automatically oscillating at the desired mode frequency. In Fig. 10b, the master resonator is the one on the left. Then the other resonator is essentially a slave resonator, which will follow the vibrations at the desired mode frequency.

The closed-loop setup can be different, in terms of the choice of master resonator. Now we discuss the following three implementation approaches.

When the master resonator is the one that is being directly driven. This implementation is as shown Fig. 10b. The drive electrode is directly applying an alternating electrostatic force to the resonator on the left, which is the master resonator. In this case, special considerations should be paid to the competition for dominance between the two modes. The underlying reason is essentially the phase condition. As can be seen from Fig. 11, for the resonator being driven, the phase of the output signal is identical for both modes; this means that if the Barkhausen phase criterion is satisfied for one mode, the oscillation phase condition of other mode is automatically satisfied. If the two modes have similar amplitude as shown in Fig. 11, then the oscillator will not work.

To circumvent this problem, the amplitudes of the two modes need to be significantly different, as shown in Fig. 12. In this case, the mode with the higher amplitude can satisfy the Barkhausen amplitude criterion, while the other mode with lower amplitude cannot.

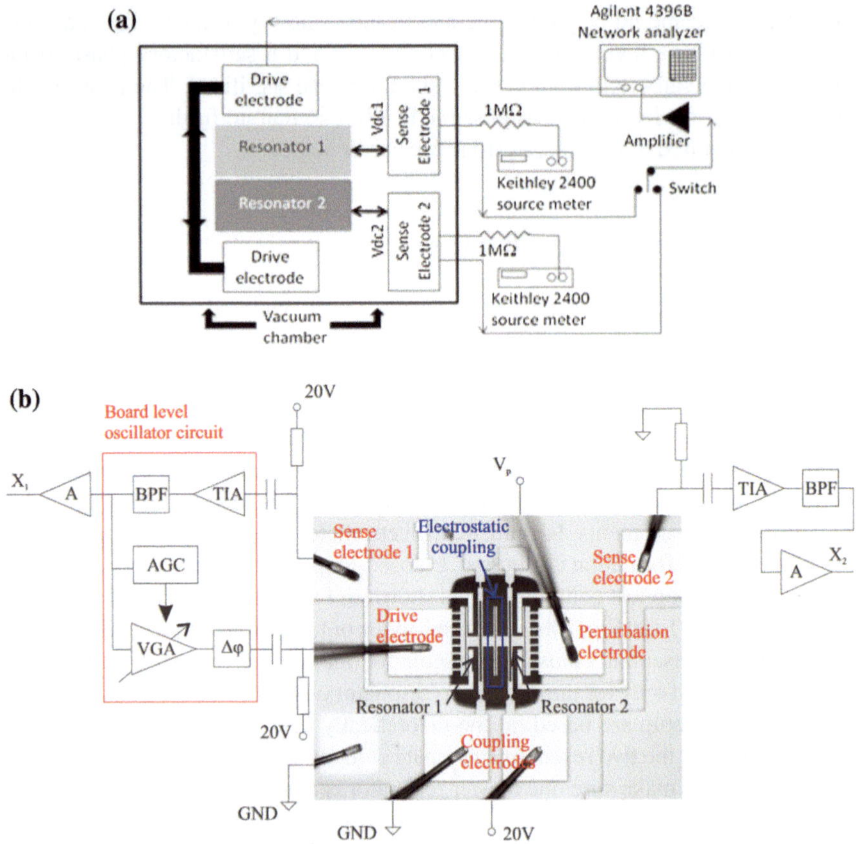

Fig. 10 Test configurations for mode-localized electrometers: **a** an open-loop test configuration [33]; **b** a closed-loop test configuration [39]

One advantage of this approach is lower oscillator phase noise, thanks to the high amplitude of the master resonator. However, as shall be discussed in Sect. 3.4, the resolution of a mode-localized electrometer depends on the operating amplitude ratio, and is optimum when the operating amplitude ratio is approximately 1.22. With this approach, as shown in Fig. 12, the operating amplitude ratio is higher than 1.22, meaning that the resolution would be higher than the optimum value.

When the master resonator is the one that is being driven through coupling. In this case, the closed-loop configuration is as shown in Fig. 13 [40]. The master resonator, which determines the frequency is the resonator on the right; whereas the slave resonator is the resonator being directly driven (the resonator on the left).

As can be observed from Fig. 11, the resonator being driven through coupling has a 180° phase difference between the two modes. If the Barkhausen phase condition of one mode is satisfied, naturally the other mode has a phase margin of 180°. Therefore, the oscillation can be very stable, free from the interference originated from the other

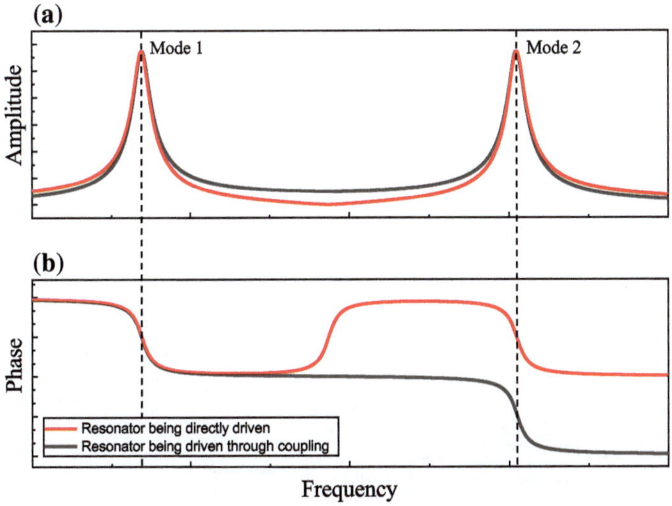

Fig. 11 Phase considerations for mode-localized electrometers in a balanced situation where $\delta k = 0$: **a** amplitude of two modes; **b** phase of two modes

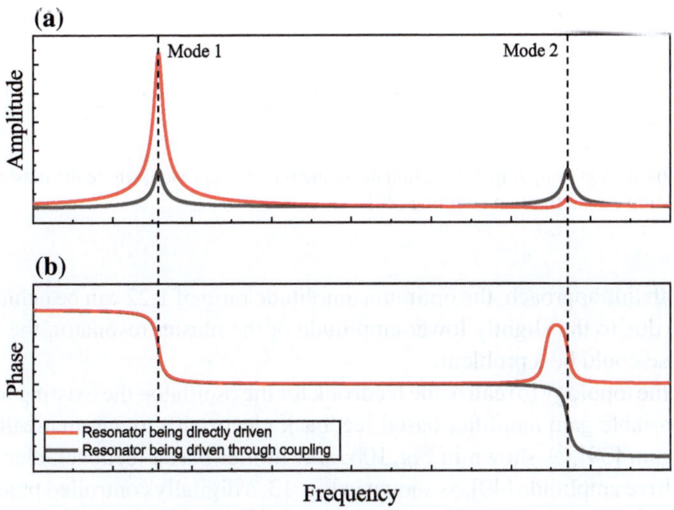

Fig. 12 Phase considerations for mode-localized electrometers in a balanced situation where $\delta k \neq 0$: **a** amplitude of two modes; **b** phase of two modes

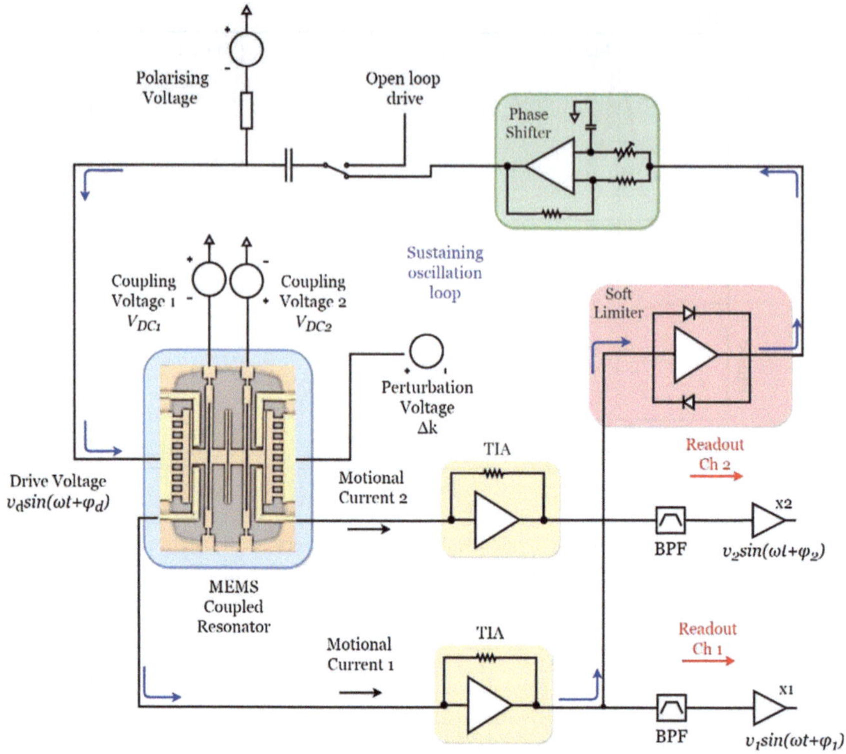

Fig. 13 Closed-loop setup implementation using the second approach, where the master resonator is the one being driven through coupling [40]

mode. With this approach, the optimum amplitude ratio of 1.22 can be achieved [18]. However, due to the slightly lower amplitude of the master resonator, the oscillator phase noise could be a problem.

As for the topology to realize the feedback for the oscillator, the existing topologies include variable gain amplifier based feedback electronics to automatically control the loop gain [39], as shown in Fig. 10b; soft-limiter based feedback electronics to limit the drive amplitude [40], as shown in Fig. 13; a digitally controlled phase-locked loop (PLL) based feedback electronics [18]. Currently, there is no strong evidence as to which feedback circuitry is the optimal for the mode-localized electrometers applications. Therefore, this can be the emphasis of further studies.

3.3 *Sensor Output Metrics*

In Sect. 2.1, it is surmised that the output metric is amplitude ratio. However, to measure the mode-localization effect in mode-localized sensors, there are a couple of additional methods, including eigenstates shift [33], amplitude difference [41]. In this section, the reason why this choice is made is explained.

From the derivations in Sect. 2.1, neglecting the effect of damping, the amplitudes of a 2DoF mode-localized electrometer at the frequency of ω_1 can be estimated as:

$$|X_1(\omega_1)| \approx \frac{1}{2}\left|\frac{k\delta k + \sqrt{(k\delta k)^2 + 4k_c^2}}{c\omega_1\sqrt{(k\delta k)^2 + 4k_c^2}}\right| |F_1| \tag{23}$$

$$|X_2(\omega_1)| \approx \left|\frac{k_c}{c\omega_1\sqrt{(k\delta k)^2 + 4k_c^2}}\right| |F_1| \tag{24}$$

First of all, it can be easily observed that the amplitude difference is proportional to the drive force, F_1. Any fluctuations in the drive force, or actuation signal, is directly transferred to the output. Therefore, the voltage difference is not a preferred output metric, despite of its high scale factor [41].

On the contrary, the output metrics based on eigenstate shift and amplitude ratio does not depend on the drive force; hence it is considered less noisy compared to voltage difference.

Historically, the first demonstrated mode-localized sensors utilize an output metric based on eigenstate shift [20]. First of all, we define the eigenstate vector of the 2DoF system as:

Definition 2

$$\text{Eigenstate} = \mathbf{u} = \begin{bmatrix} \frac{X_1}{\sqrt{X_1^2+X_2^2}} \\ \frac{X_2}{\sqrt{X_1^2+X_2^2}} \end{bmatrix} \tag{25}$$

The eigenstate shift output metric is defined as [33]:

Definition 3

$$\text{Eigenstate shift} = \mathbf{\Delta u} = \frac{|\mathbf{u_1} - \mathbf{u_0}|}{|\mathbf{u_0}|} \tag{26}$$

Where $\mathbf{u_1}$ is the eigenstate vector after perturbation, and $\mathbf{u_0}$ is the eigenstate vector before perturbation.

Based on these definitions, and the calculations of amplitude ratios described in Sect. 2.1, the two output metric can be compared directly in terms of its scale factor. The results are shown in Fig. 14.

From the figures, it can be obtained that the scale factor of eigenstate shift is approximately, for $\delta k \ll k_c$:

$$\mathfrak{C}_\mathbf{u} \approx \frac{k}{4k_c} \tag{27}$$

As the perturbation increases, the scale factor decreases. In comparison, in the same operating region, where $\delta k \ll k_c$, the scale factor of amplitude ratio is twice

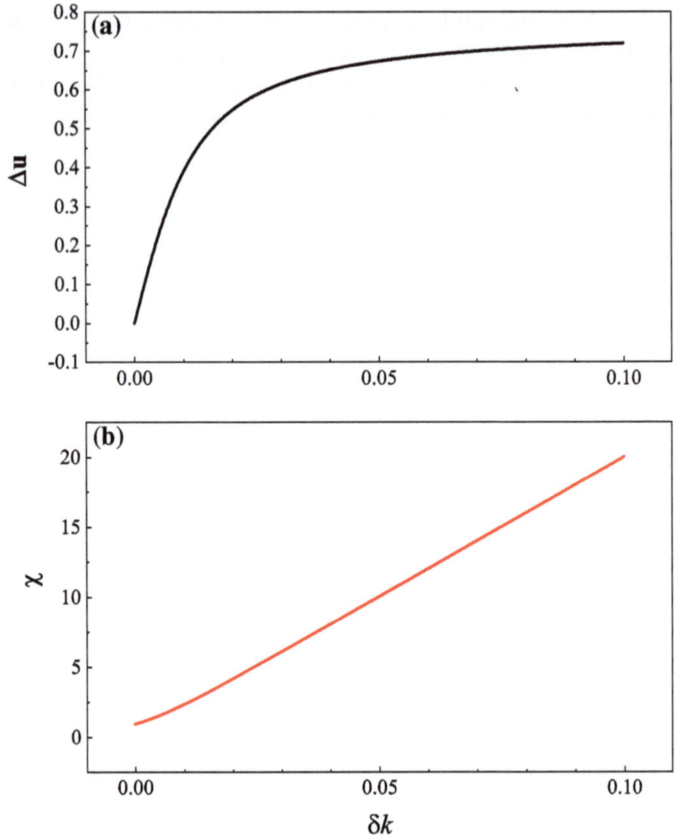

Fig. 14 Output metric comparisons for a 2DoF mode-localized electrometers: **a** eigenstate shift with respect to stiffness perturbation, **b** amplitude ratio change with respect to stiffness perturbation

of that of eigenstate shift; for large perturbations, the scale factor of amplitude ratio is much higher than that of eigenstate shift.

In addition to scale factor enhancement, the hardware to calculate amplitude ratio can be easily implemented. The oscillating output of the resonators are ideally fed into a two-channel low-noise lock-in amplifier, which can then be demodulated to extract the amplitudes. Then the amplitude ratio can be calculated using an offline calculation program, which is the common approach concurrently. However, the amplitude ratios can also be calculated using a real-time ratiometric function, which should be relatively straightforward to implement in digital domain.

To conclude, the amplitude ratio is a preferred output metric for mode-localized electrometers.

3.4 Sensor Noise Floor

From the derivations in Sect. 3.3, the magnitude of amplitude ratio, $|\chi(\omega_1)|$, can be expressed as:

$$|\chi(\omega_1)| = \left|\frac{X_1(\omega_1)}{X_2(\omega_1)}\right| \approx \left|\frac{k\delta k + \sqrt{(k\delta k)^2 + 4k_c^2}}{2k_c}\right| \tag{28}$$

The following can then be derived:

$$\sqrt{(k\delta k)^2 + 4k_c^2} = \left(\frac{[\chi(\omega_1)]^2 + 1}{\chi(\omega_1)}\right) k_c \tag{29}$$

Substituting Eq. (29) into Eq. (24), the amplitude of resonator 2 can be estimated by:

$$|X_2(\omega_1)| \approx \left|\frac{Q}{k + k_c} \frac{\chi(\omega_1)}{[\chi(\omega_1)]^2 + 1}\right| |F_1| \tag{30}$$

Further, substituting Eq. (30) into Eq. (28), the amplitude of resonator 1 can be obtained:

$$|X_1(\omega_1)| \approx \left|\frac{Q}{k + k_c} \frac{[\chi(\omega_1)]^2}{[\chi(\omega_1)]^2 + 1}\right| |F_1| \tag{31}$$

Additionally, using Eqs. (28) and (29), the scale factor with respect to δk, as a function of the amplitude ratio, can be written down as:

$$\mathcal{C}_\chi(\omega_1) \approx \left|\frac{[\chi(\omega_1)]^2}{[\chi(\omega_1)]^2 + 1} \frac{k}{k_c}\right| \tag{32}$$

Currently, the resolution of the majority of the existing mode-localized sensors, including mode-localized electrometers, is limited by the electronics noise originated from the front-end TIA. Hence, it is fair to assume that the dominant noise source is the front-end TIA. In this case, the two channel front-end TIA, as shown in Fig. 13, are independence noise sources, meaning that the noise correlation between the two channels is zero. Based on these assumptions, we can continue to derive the resolution of the mode-localized electrometersfurther. The variance of the amplitude ratio (or the noise power), $\sigma^2(X_1/X_2)$, is given by Pandit et al. [40]:

$$\sigma^2(X_1/X_2) = \frac{[\mu(X_1)]^2}{[\mu(X_2)]^2}\left[\frac{\sigma^2(X_1)}{[\mu(X_1)]^2} + \frac{\sigma^2(X_2)}{[\mu(X_2)]^2}\right] \tag{33}$$

where $\mu(X_i) = E(X_i)$ $(i = 1, 2)$ is the expected value of the ith resonator's amplitude, which is given by Eqs. (31) and (30), respectively. $\sigma^2(X_i)$ $(i = 1, 2)$ is the noise power of the ith resonator's amplitude, whose main contributor is the noise of front-end TIA. The NSD is given by Eq. (22). And the noise power of each amplitude is essentially the total noise, which can be calculated by integrating the NSD within the measurement bandwidth. So here we assume it is a constant value given by $\sigma^2(X_e)$, where the subscript denotes the origin of the noise is from electronics. Substituting $\sigma^2(X_i) = \sigma^2(X_e)$ $(i = 1,\ 2)$, and the expression of $\mu(X_i)$ into Eq. (33), we can obtain:

$$\sigma^2(X_1/X_2) = \sigma^2(X_e)\left|\frac{k+k_c}{Q}\right|^2 \frac{\left\{[\chi(\omega_1)]^2+1\right\}^3}{[\chi(\omega_1)]^2} \tag{34}$$

The input-referred noise power can be estimated using the scale factor:

$$\sigma^2(\delta k) = \frac{\sigma^2(X_1/X_2)}{[\mathcal{C}_\chi(\omega_1)]^2} = \sigma^2(X_e)\left|\frac{(k+k_c)k_c}{kQ}\right|^2 \frac{\left\{[\chi(\omega_1)]^2+1\right\}^5}{[\chi(\omega_1)]^6} \tag{35}$$

Assuming weak coupling where $k_c \ll k$, the input-referred NSD can then be estimated as:

$$\langle \delta k(f)\rangle = \sqrt{\frac{\mathrm{d}\sigma^2(\delta k)}{\mathrm{d}f}} = \langle X_e(f)\rangle\left|\frac{k_c}{Q}\right|\sqrt{\frac{\left\{[\chi(\omega_1)]^2+1\right\}^5}{[\chi(\omega_1)]^6}} \tag{36}$$

Where $\langle \delta k(f)\rangle$ and $\langle X_e(f)\rangle = \sqrt{\frac{\mathrm{d}\sigma^2(X_e)}{\mathrm{d}f}}$ is the frequency dependent NSD of the input-referred stiffness perturbation and front-end TIA, respectively. From Eq. (36), it can be observed that the input-referred NSD depends critically on the amplitude ratio $\chi(\omega_1)$. Also can be observed is the fact that there is a minimum NSD value that can be achieved. Using fundamental calculus, it can be found that the minimum NSD can be achieved at $\chi(\omega_1) = \sqrt{3/2} \approx 1.22$.

It should be pointed out that the scale factor for the mode-localized electrometers is intrinsically nonlinear, and an optimum operating point for the best scale factor linearity is at amplitude ratios greater than 5. On top of that, the larger operating amplitude ratio, the smaller the nonlinearity. However, as shown in Fig. 15a and in Eq. (36), the input-referred noise floor deteriorates as the amplitude ratio increases, and has a second order dependency on the amplitude ratio. This indicates that there is a trade-off between the input-referred noise floor and scale factor nonlinearity, as shown in Fig. 15b. For high resolution electrometers, it is recommended to operate near the amplitude ratio of 1.22, in spite of a worse linearity.

It should be pointed out that the resolution is dependent on the amplitude ratio, thus the input stimulus. Therefore, in order to maintain a high resolution sensing, it is necessary to maintain the amplitude ratio at this lowest point. To realize this, a closed-loop feedback control system could be sufficient.

Through these resolution optimization efforts, a minimum noise floor of 16ppb/$\mathrm{Hz}^{1/2}$ (between 1 and 10 Hz) in terms of δk can be achieved [18]. Using the dimensions mentioned in the paper, $d = 2\,\mu\mathrm{m}$, $L = 300\,\mu\mathrm{m}$, $k = 1000\,\mathrm{N/m}$, $V_0 = 24\,\mathrm{V}$,

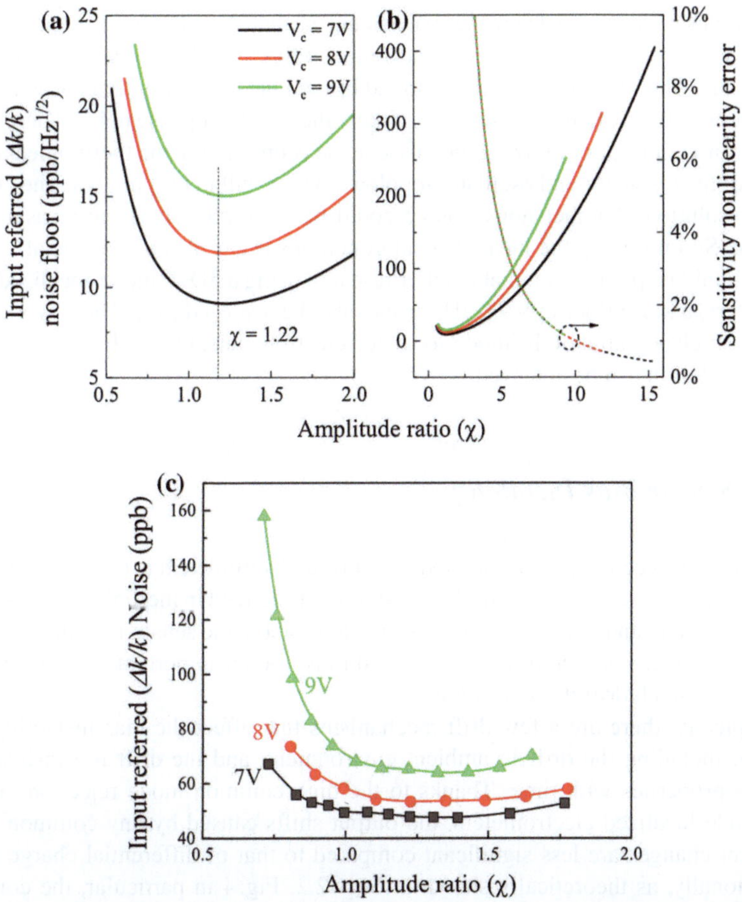

Fig. 15 Input-referred noise floor versus amplitude ratio [18]: **a** showing the theoretical optimal operating amplitude ratio for the best input-referred noise floor is 1.22, **b** showing a trade-off between input-referred noise floor and scale factor nonlinearity, **c** showing the measured optimal input-referred noise floor versus amplitude ratio

if the electrons are applied axially, the estimated noise floor is approximately 0.082 fC/Hz$^{1/2}$ or 510 e/Hz$^{1/2}$ using Eq. (19). If the electrostatic force caused by charge injection is parallel to the direction of vibration (approach 2), using Eq. (21), an estimated noise floor of 0.0013 fC/Hz$^{1/2}$ or 8.3 e/Hz$^{1/2}$, could potentially be achieved.

It should be pointed out that the work is based on the front-end TIA shown in Fig. 9. And the results are obtained using a mode-localized device which has a lot of room for improvement in its design. For instance, the Q-factor can be improved using slotted DETF design [31]. A systematic optimization approach, combining the resonator and electronics, can help further improving the noise floor of the sensor.

Another observation from the experiment is that the measured output PSD is approximately twice of the calculated output PSD, based on measured front-end TIA NSD. This indicates that there are additional noise contributions. It is thought that the one of the main noise sources is likely the oscillator phase noise. Therefore, a further direction for optimizing the phase noise is optimizing the thermomechanical noise of the resonator and oscillator topology, which will in turn improve the closed-loop resolution. Another noise source could be the the noise of the measurement system. So improving the measurement system noise can also be beneficial.

It should be pointed out that another research using a 3DoF mode-localized electrometers, a noise floor of 9.21 e/Hz$^{1/2}$ can also be achieved [36]. These work show that the high potential of the mode-localized electrometers, that high scale factor and high resolution can be achieved.

3.5 Sensor Bias Instability

Another key specification of the mode-localized electrometers is the bias instability of the sensor. Conventionally, the bias instability is defined in inertial sensors, such as accelerometers and gyroscopes, to describe how stable the sensor is within a certain period of time. For electrometers, it should also be an important specification, to show how much drift the sensor has.

Typically, there are a few drift mechanisms that affect the bias instability of a sensor, including the drift in ambient environment, and the drift in sensitive element's properties with time. Thanks to the high common mode rejection ratio of the mode-localized electrometers, the output shifts caused by any common mode ambient changes are less significant compared to that of differential charge input. Additionally, as theoretically shown in Sect. 2.2, Fig. 4 in particular, the common mode drift in the resonators, such as intrinsic drift in the stiffness, in the mode-localized electrometers can also be eliminated.

On top of those drift mechanisms mentioned above, drifts in the electronics also contribute to the total bias instability. As shown in the derivations in Sect. 2.2, amplitude ratio is independent on the drive amplitude. This means that any drift in the drive amplitude, which can be caused by the drift in the feedback gain, can be eliminated by using the amplitude ratio output metric of mode-localized electrometers. Furthermore, the common mode drift in the transduction from resonator displacement to the output voltage can be reduced through the ratiometric readout. Last but not least, since the amplitude readout can be based on a demodulation approach using a low noise lock-in amplifier. Hence, any drifts in the DC offset, as well as low frequency $1/f$ noise, within the front-end readout circuitry can be nullified.

These claims made above are evidenced by two separate studies on a generic mode-localized sensor [40] and a mode-localized accelerometer [34]. In both devices, it can be observed that the mode-localized sensor and amplitude ratio readout provide

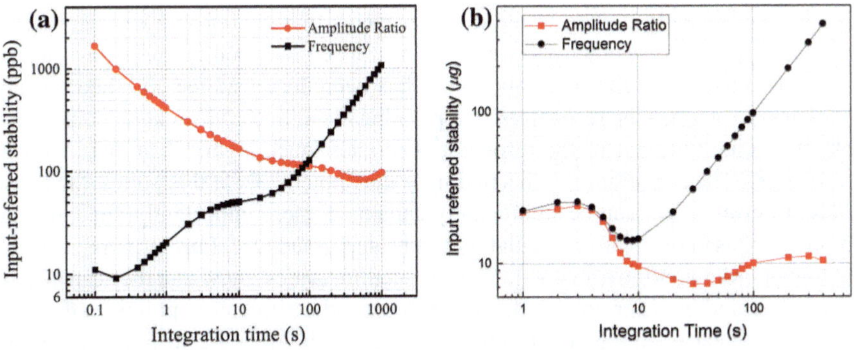

Fig. 16 Input-referred bias instability of mode-localized sensors: **a** a generic mode-localized sensor [40]; **b** a mode-localized accelerometer [34]

a superior bias instability for long integration times, as compared to frequency shift readout. This indicates that the mode-localized sensor and its output metrics are more stable for long term measurements (Fig. 16).

4 Further Methods to Improve the Sensor

As mentioned above, the input-referred noise floor of mode-localized electrometers can be further improved through optimizations in resonator design (i.e. on topology and dimension), coupled resonator system design (i.e. number of DoF and coupling mechanism), as well as electronics design (i.e. front-end TIA and oscillator noise optimization). In the meantime, studies on the methods to improve the sensor performance also exist in the literature. Here, we summarize some of the related work that could be potentially useful for future research in mode-localized electrometers.

4.1 Utilizing Nonlinearities to Improve Resolution

All the resonators mentioned in the chapter up till this point are assumed to be operating in linear regimes. Hence, the transfer function based analysis can be applied for the purpose of analysis. And this simplicity for analysis was actually one of the main reasons why only the linear resonators were considered. However, recent advances in MEMS and NEMS resonator experimental techniques, as well as our increasingly deepening understanding on the nonlinear mechanics (thanks to the pioneers such as Prof. Michael Roukes at Caltech, Prof. Thomas Kenny at Stanford, Prof. Steven Shaw at Florida Tech and many more), have enabled the community to closely examine the effect of nonlinearities within micro/nano-scale resonators.

It has been recently demonstrated in experiments that nonlinear micro/nano-scale resonators can be very precisely controlled in a closed-loop configuration, by varying the feedback phase [42]. Not only that, it has been demonstrated and theoretically explained that when the resonators are operating at nonlinear bifurcation points,[5] the effective quality factor of the resonator[6] becomes much higher than that of linear resonators. The importance of this finding is that, potentially at this points, the phase noise filtering effect can be significantly enhanced; more importantly, fundamental limits to phase noise within oscillators can be surpassed [43]. Furthermore, analysis of higher order nonlinearities, e.g. 5th order nonlinearities, show that they can also be controlled and used for noise reduction purposes; the bifurcation points on the high order nonlinearity can also help minimize the amplitude noise to frequency noise conversion. These studies pave new paths for utilizing the nonlinearity to enhance the resolution of sensors based on resonator structures.

Following these studies on nonlinearities within a single resonator, it has been proposed that the nonlinearities can also be employed on mode-localized sensors, and in the future in mode-localized electrometers.

In the first work [44], it has been demonstrated that 3rd order nonlinearities can be utilized in a generalized mode-localized sensor. Through analysis and measurement, it has been shown that at a top bifurcation point (shown in Fig. 17), the slope of the amplitude ratio derivative with respect to the phase derivative becomes close to zero.

This indicates that any phase fluctuations within an oscillator for mode-localized sensors do not affect the amplitude ratios. Additionally, at the top bifurcation point, the amplitude of the resonators are above the critical linear amplitude, meaning that the signal-to-noise ratio is also improved. Combining these two effects, the output amplitude ratio noise floor can be improved, as shown in Fig. 18. As can be observed from Fig. 18, both bifurcation points are useful for improving the noise floor. More details on this study can be found in [44].

In another study, the noise reduction capability of high order nonlinearities, i.e. up to 5th order mechanical and electrical nonlinearities within a coupled resonator, are examined [45]. For 5th order nonlinear system described by a double hysteresis response, 4 bifurcation points should occur on the resonators' frequency response. In this study, a closed-loop control system is realized, and the closed-loop responses are measured. The phase and signal gain of resonators 1 and 2 within a 2DoF mode-localized sensoris shown in Fig. 19,[7] and a double hysteresis behaviour can be observed.

The effect of 3 out of 4 bifurcation points on stability enhancement are discussed, and an optimum bifurcation point in terms of amplitude stability is proposed for amplitude based readout metrics (see Fig. 20). The reason for the high stability is attributed to both high signal-to-noise ratio at this point and its near zero slope of the amplitude derivative with respect to the phase derivative.

[5]The unstable region in the frequency response of the resonators, above the critical linear amplitude.

[6]Also, the slope of derivative of phase with respect to frequency, $d\phi/df$.

[7]In this study [45], resonator 1 and 2 are defined as the resonator being directly driven by an AC drive voltage, and the resonator being driven through coupling, respectively.

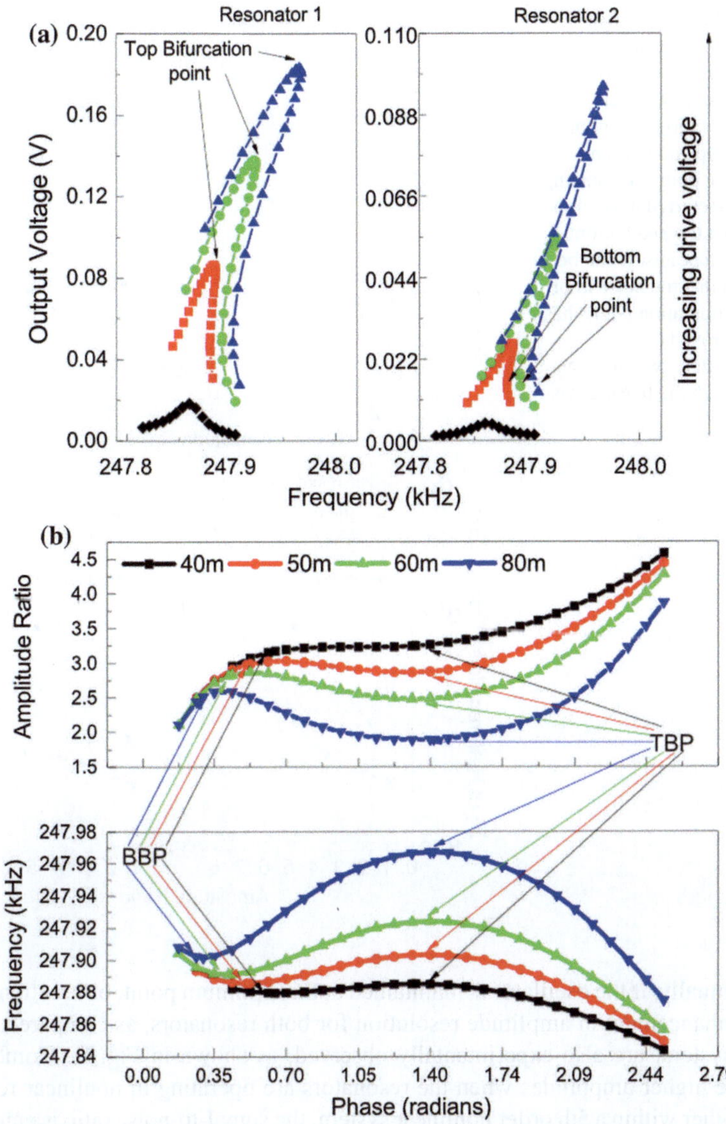

Fig. 17 Measured nonlinear responses of a 2DoF mode-localized sensor [44]: **a** showing the individual frequency responses of resonator 1 and 2, the top bifurcation point and bottom bifurcation point are defined in the figure; **b** showing the amplitude ratio and frequency with respect to closed-loop phase, special attention should be paid to the two bifurcation points, where both the slope of amplitude ratios derivative and frequency derivative with respect to phase derivative is zero

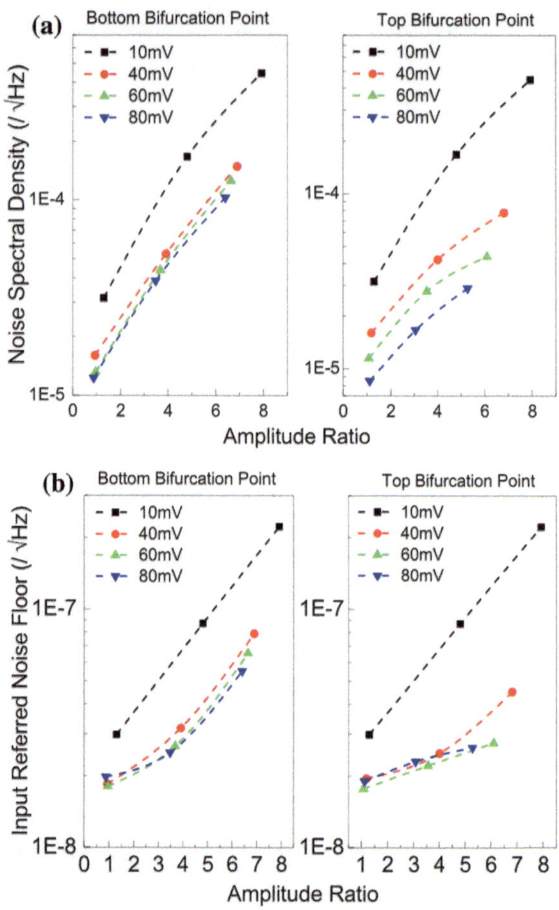

Fig. 18 Measured noise floor of a nonlinear 2DoF mode-localized sensor [44] at 30 Hz: **a** showing the amplitude ratio noise floor at both the top and bottom bifurcation points; **b** showing the input-referred noise floor at both the top and bottom bifurcation points. It can be observed that the noise floor at both bifurcation points has been improved by approximately two times as compared to the linear cases

Eventually, if the oscillator is maintained at this optimum point, bifurcation point 2, an enhancement in amplitude resolution for both resonators, as compared to the linear system, are also experimentally observed, as shown in Fig. 21. Combining with the higher amplitudes when the resonators are operating in nonlinear regime, even higher within a 5th order nonlinear system, the signal-to-noise ratio is enhanced by over 30 dB. As discussed in Sect. 3.4, the output noise floor of mode-localized electrometers depends critically on the amplitude signal-to-noise ratio, hence, this approach can be useful to further improve the noise floor.

However, it has been demonstrated that a reduction in scale factor could occur at these points, the scale factor at the bifurcation point 2 within a 5th order nonlinear 2DoF system should be characterized in the future.

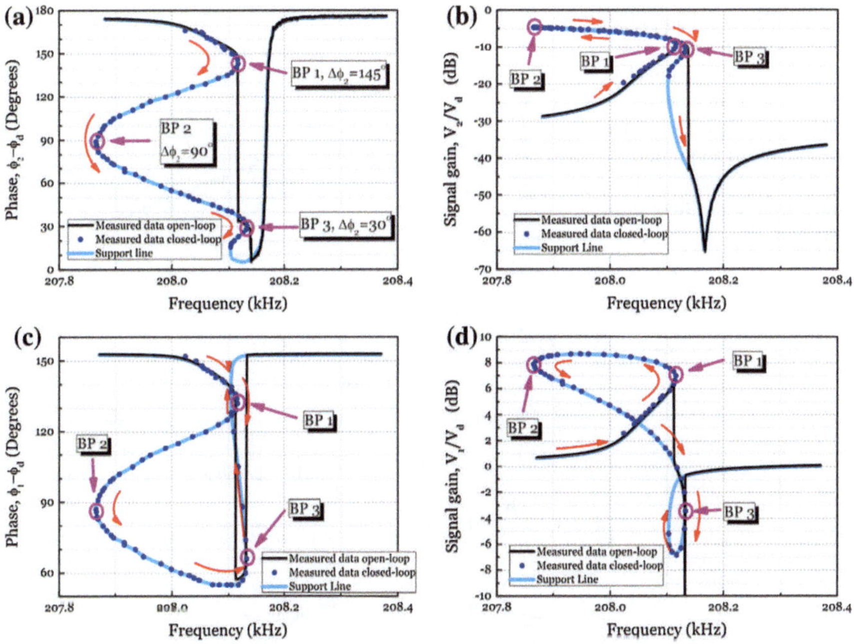

Fig. 19 Measured phase and signal gain of a 5th order nonlinear 2DoF mode-localized sensor [45]: **a** the phase response of resonator 2; **b** the signal gain of resonator 2; **c** the phase response of resonator 1; **d** the signal gain of resonator 1. The definitions of bifurcation points are given in the figures

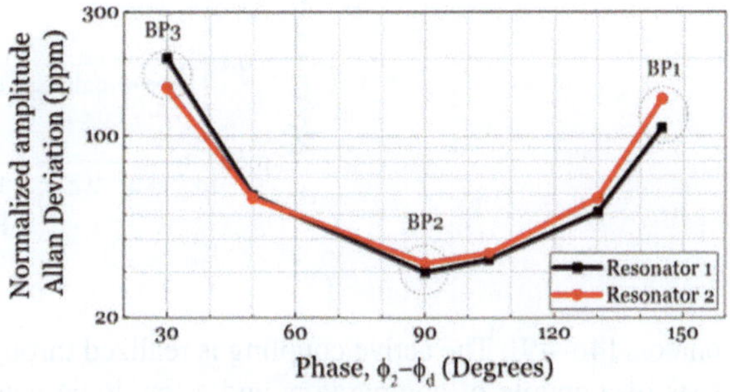

Fig. 20 Measured amplitude stability of the 5th order nonlinear 2DoF system at integration time $\tau = 0.1$ s (representing short-term stability) with respect to the operating points [45]

4.2 *Employing Alternative Coupling Methods*

The mode-localized electrometers and mode-localized sensors discussed in the previous sections have used either a mechanical coupling or an electrical coupling. These two methods are simply to implement and easy to design. However, despite of the obvious advantages of these coupling elements, whether these coupling designs are the optimum is still uncertain.

In a series of recently published papers, a group of scientists lead by Prof. Jérôme Juillard have proposed a new active coupling approach between two identical res-

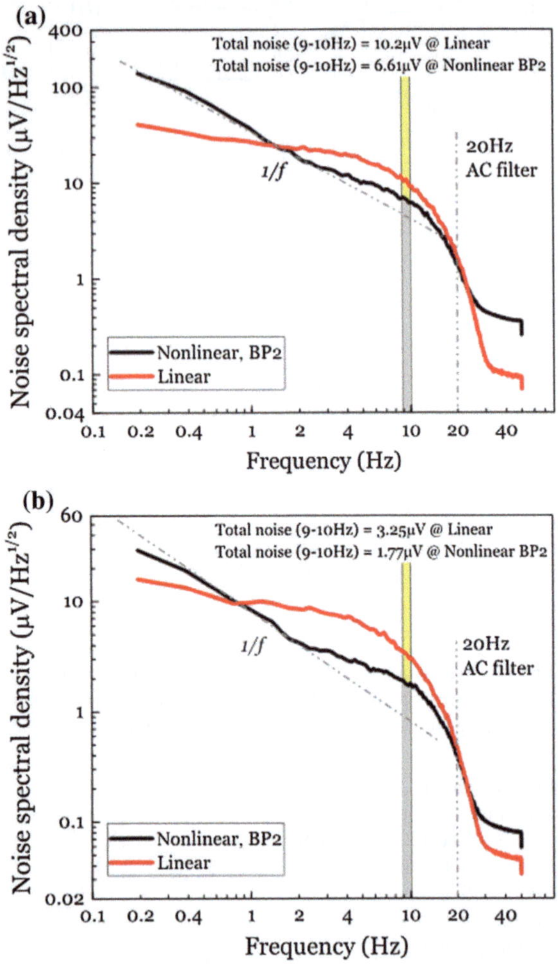

Fig. 21 Extracted amplitude noise floor of the 5th order nonlinear 2DoF system at bifurcation point 2, as compared to that of the linear system [45]: **a** amplitude noise floor of resonator 1; **b** amplitude noise floor of resonator 2

onators [46–49]. The active coupling is realized through a digital mixer, which consists of a couple of comparators and a few logic gates, and is implemented using CMOS technology [48]. The schematic of this approach is shown in Fig. 22. The advantage of this active coupling is that it can reach potentially very low coupling strength, i.e. ($k_c Q/k \approx 1$) [49]. This recently proposed approach is termed as *mutually injection locked oscillator* (MILO).

Unlike the mode-localized sensors where the resonators are close to each other, in the MILO configuration proposed, two identical CMOS-MEMS resonators are fabricated on a same chip, but are placed away from each other. This shows a potential advantage of the topology, that two oscillators can potentially be implemented individually; thus potentially the oscillators does not necessarily need to be of the same type, as long as the resonant frequencies are identical.

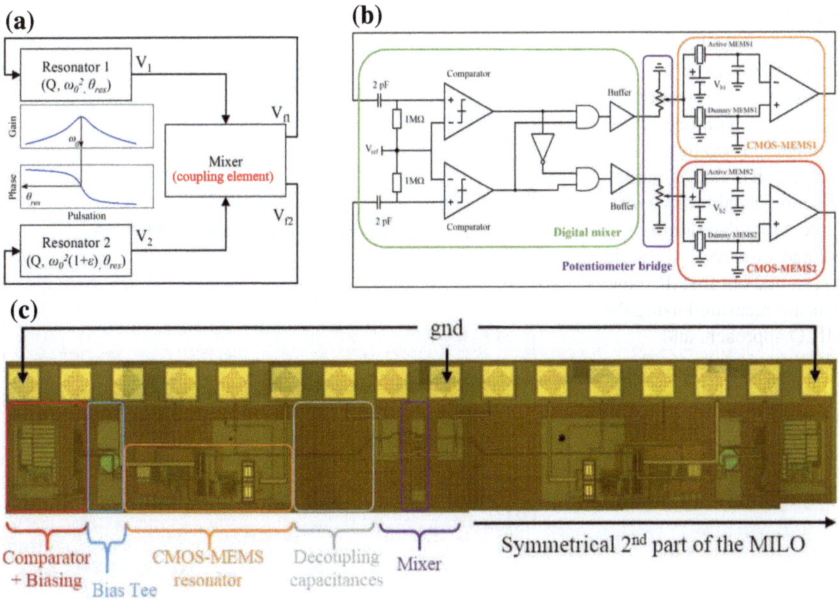

Fig. 22 Schematic of the MILO sensor [48]: **a** high level schematic of MILO, showing the mixer as the active coupling element; **b** detailed schematic of the MILO implementation using CMOS-MEMS; **c** optical graph of the physical implementation of MILO based on CMOS-MEMS technology

The operation of this new approach is predominantly based on the phase readout. When the two resonators are identical, with the active coupling, the two oscillators are injection-locked with each other; and the phase difference between them is initially $\pi/2$. With the introduction of a small stiffness mismatch, which can be induced by an electron charge injection, the two oscillators are still injection-locked; but the phase difference between the two resonators would shift away from $\pi/2$. And the nominal scale factor is given by Juillard et al. [49]:

$$\mathcal{C}_{\phi} = \frac{k^2}{Qk_c^2} \tag{37}$$

Due to the high scale factor, and high measurement resolution on phase, it has been demonstrated in theory that the MILO can achieve higher resolution than mode-localized electrometers.

Additionally, similar to the conventional mode-localized sensors with passive coupling (as opposed to active coupling in MILO approach), the MILO sensors also provide excellent common mode rejection [47]. This is shown in Fig. 23. As can be observed from Fig. 23, the phase readout is independent to either the common mode ambient temperature changes or polarization voltage variations.

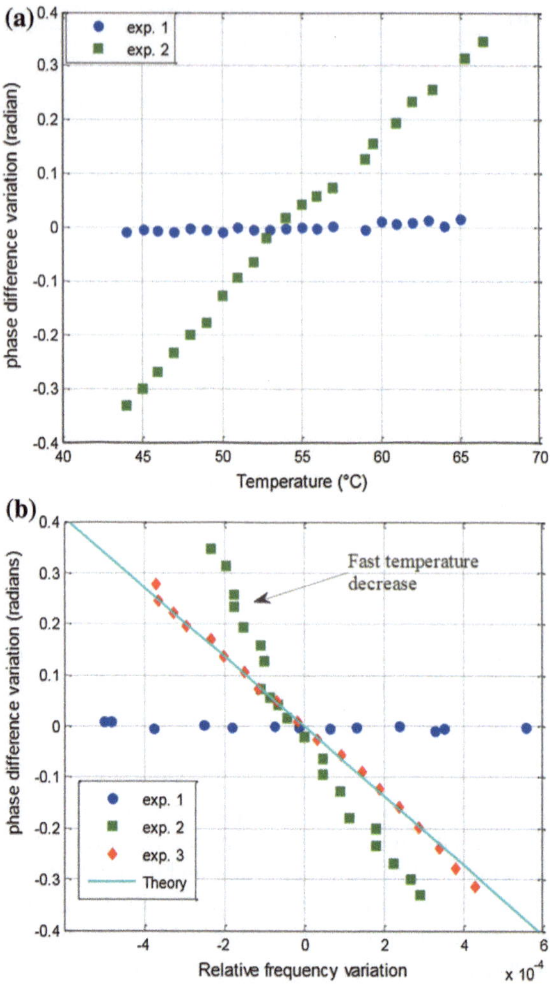

Fig. 23 Common mode rejection capability of the MILO sensor [47]: **a** common mode rejection capabilities with respect to ambient temperature changes; **b** common mode rejection capabilities with respect to polarization voltage variations. It should be pointed out that the blue dots are measured using the MILO approach, and excellent common mode rejection capability is demonstrated

However, one downside of the approach is that the two resonant frequencies need to be sufficiently close to each other, so that the two oscillators can be mutually injection locked. Despite of the fact that the initial mismatch can be tuned by a DC voltage; however, during the operation, large stiffness mismatch caused by a large charge injection can potentially dislodge the injection lock, making the resonators malfunction. It has been derived that the locking range is inversely proportional to the Q-factor of the resonators. A same dependence of the scale factor on Q-factor can be seen from Eq. (37). Hence, there is a trade-off between the scale factor and locking range. To enhance the locking range, one method based on nonlinearity is proposed [50].

Nevertheless, these studies show that alternative coupling approaches can be proposed to enhance the performance of mode-localized electrometers.

5 Conclusions

In this chapter, a new approach for micro/nano-scale electrometers based on mode-localization effect is proposed and discussed. The operation principle of mode-localized electrometers is presented, and the advantages such as high scale factor and high common mode rejection capability are also analyzed. Furthermore, the design challenges, including low noise front-end electronics design and noise floor optimization approaches of the mode-localized electrometers are also discussed. Finally, a few new directions to further enhance the sensor performance are also drawn out from the literature [25].

References

1. A.N. Cleland, M.L. Roukes, A nanometre-scale mechanical electrometer. Nature **392**(6672), 160 (1998)
2. J.E.Y. Lee, B. Bahreyni, A.A. Seshia, An axial strain modulated double-ended tuning fork electrometer. Sens. Actuators Phys. **148**(2), 395–400 (2008)
3. A.A. Seshia, M. Palaniapan, T.A. Roessig, R.T. Howe, R.W. Gooch, T.R. Schimert, S. Montague, A vacuum packaged surface micromachined resonant accelerometer. J. Microelectromech. Syst. **11**(6), 784–793 (2002)
4. R.H. Olsson, K.E. Wojciechowski, M.S. Baker, M.R. Tuck, J.G. Fleming, Post-CMOS-compatible aluminum nitride resonant MEMS accelerometers. J. Microelectromech. Syst. **18**(3), 671–678 (2009)
5. C. Comi, A. Corigliano, G. Langfelder, A. Longoni, A. Tocchio, B. Simoni, A resonant microaccelerometer with high sensitivity operating in an oscillating circuit. J. Microelectromech. Syst. **19**(5), 1140–1152 (2010)
6. X. Zou, P. Thiruvenkatanathan, A.A. Seshia, A seismic-grade resonant MEMS accelerometer. J. Microelectromech. Syst. **23**(4), 768–770 (2014)
7. D.D. Shin, C.H. Ahn, Y. Chen, D.L. Christensen, I.B. Flader, T.W. Kenny, Environmentally robust differential resonant accelerometer in a wafer-scale encapsulation process, in *2017 IEEE 30th International Conference on Micro Electro Mechanical Systems (MEMS)* (IEEE, 2017), pp. 17–20
8. Y. Yin, Z. Fang, F. Han, B. Yan, J. Dong, Q. Wu, Design and test of a micromachined resonant accelerometer with high scale factor and low noise. Sens. Actuators Phys. **268**, 52–60 (2017)
9. R. Sunier, T. Vancura, Y. Li, K.U. Kirstein, H. Baltes, O. Brand, Resonant magnetic field sensor with frequency output. J. Microelectromech. Syst. **15**(5), 1098–1107 (2006)
10. B. Bahreyni, C. Shafai, A resonant micromachined magnetic field sensor. IEEE Sens. J. **7**(9), 1326–1334 (2007)
11. M. Li, S. Nitzan, D.A. Horsley, Frequency-modulated Lorentz force magnetometer with enhanced sensitivity via mechanical amplification. IEEE Electron Device Lett. **36**(1), 62–64 (2015)
12. Y. Hui, T. Nan, N.X. Sun, M. Rinaldi, High resolution magnetometer based on a high frequency magnetoelectric MEMS-CMOS oscillator. J. Microelectromech. Syst. **24**(1), 134–143 (2015)
13. T.R. Albrecht, P. Grütter, D. Horne, D. Rugar, Frequency modulation detection using high-Q cantilevers for enhanced force microscope sensitivity. J. Appl. Phys. **69**(2), 668–673 (1991)
14. F.J. Giessibl, S. Hembacher, H. Bielefeldt, J. Mannhart, Subatomic features on the silicon (111)-(7Œ7) surface observed by atomic force microscopy. Science **289**(5478), 422–425 (2000)
15. J.A. Harley, T.W. Kenny, A high-stiffness axial resonant probe for atomic force microscopy. J. Microelectromech. Syst. **10**(3), 434–441 (2001)

16. E. Algre, Z. Xiong, M. Faucher, B. Walter, L. Buchaillot, B. Legrand, MEMS ring resonators for laserless AFM with sub-nanoNewton force resolution. J. Microelectromech. Syst. **21**(2), 385–397 (2012)
17. Y. Zhu, J.E.Y. Lee, A.A. Seshia, A resonant micromachined electrostatic charge sensor. IEEE Sens. J. **8**(9), 1499–1505 (2008)
18. C. Zhao, M. Pandit, G. Sobreviela, A. Mustafazade, S. Du, X. Zou, A. Seshia, On the noise optimization of resonant MEMS sensors utilizing vibration mode localization. Appl. Phys. Lett. **112**(19), 194103 (2018)
19. P. Thiruvenkatanathan, J. Yan, J. Woodhouse, A. Aziz, A.A. Seshia, Effects of mechanical and electrical coupling on the parametric sensitivity of mode localized sensors, in *2009 IEEE International Ultrasonics Symposium* (IEEE, 2009), pp. 1183–1186
20. M. Spletzer, A. Raman, A.Q. Wu, X. Xu, R. Reifenberger, Ultrasensitive mass sensing using mode localization in coupled microcantilevers. Appl. Phys. Lett. **88**(25), 254102 (2006)
21. P. Thiruvenkatanathan, J. Yan, A.A. Seshia, Differential amplification of structural perturbations in weakly coupled MEMS resonators. IEEE Trans. Ultrason. Ferroelectr. Freq. Control **57**(3), 690–697 (2010)
22. M. Pandit, C. Zhao, G. Sobreviela, A.A. Seshia, Immunity to Temperature Fluctuations in Weakly Coupled MEMS Resonators, in *2018 IEEE Sensors* (IEEE, 2018), pp. 1–4
23. C. Zhao, A MEMS sensor for stiffness change sensing applications based on three weakly coupled resonators, Doctoral dissertation, University of Southampton, 2016
24. H. Kang, J. Yang, H. Chang, A closed-loop accelerometer based on three degree-of-freedom weakly coupled resonator with self-elimination of feedthrough signal. IEEE Sens. J. **18**(10), 3960–3967 (2018)
25. C. Zhao, G.S. Wood, J. Xie, H. Chang, S.H. Pu, M. Kraft, A three degree-of-freedom weakly coupled resonator sensor with enhanced stiffness sensitivity. J. Microelectromech. Syst. **25**(1), 38–51 (2016)
26. H. Kang, J. Yang, H. Chang, A mode-localized accelerometer based on four degree-of-freedom weakly coupled resonators, in *2018 IEEE Micro Electro Mechanical Systems (MEMS)* (IEEE, 2018), pp. 960–963
27. B. Peng, K.M. Hu, L. Shao, H. Yan, L. Li, X. Wei, W.M. Zhang, A sensitivity tunable accelerometer based on series-parallel electromechanically coupled resonators using mode localization. J. Microelectromech. Syst. (2019)
28. M. Spletzer, A. Raman, H. Sumali, J.P. Sullivan, Highly sensitive mass detection and identification using vibration localization in coupled microcantilever arrays. Appl. Phys. Lett. **92**(11), 114102 (2008)
29. P. Thiruvenkatanathan, J. Yan, A.A. Seshia, Ultrasensitive mode-localized micromechanical electrometer, in *2010 IEEE International Frequency Control Symposium* (IEEE, 2010), pp. 91–96
30. P. Thiruvenkatanathan, J. Yan, J. Woodhouse, A. Aziz, A.A. Seshia, Ultrasensitive mode-localized mass sensor with electrically tunable parametric sensitivity. Appl. Phys. Lett. **96**(8), 081913 (2010)
31. R.N. Candler, A. Duwel, M. Varghese, S.A. Chandorkar, M.A. Hopcroft, W.T. Park, T.W. Kenny, Impact of geometry on thermoelastic dissipation in micromechanical resonant beams. J. Microelectromech. Syst. **15**(4), 927–934 (2006)
32. V. Kaajakari, T. Mattila, A. Oja, H.A.S.H. Seppa, Nonlinear limits for single-crystal silicon microresonators. J. Microelectromech. Syst. **13**(5), 715–724 (2004)
33. P. Thiruvenkatanathan, J. Yan, J. Woodhouse, A.A. Seshia, Enhancing parametric sensitivity in electrically coupled MEMS resonators. J. Microelectromech. Syst. **18**(5), 1077–1086 (2009)
34. M. Pandit, C. Zhao, G. Sobreviela, A. Mustafazade, X. Zou, A.A. Seshia, A mode-localized MEMS accelerometer with $7\mu g$ bias stability, in *2018 IEEE Micro Electro Mechanical Systems (MEMS)* (IEEE, 2018), pp. 968–971
35. H. Zhang, W. Yuan, J. Huang, B. Li, H. Chang, A high-sensitive resonant electrometer based on mode localization of the weakly coupled resonators, in *2016 IEEE 29th International Conference on Micro Electro Mechanical Systems (MEMS)* (IEEE, 2016), pp. 87–90

36. J. Yang, H. Kang, H. Chang, A micro resonant electrometer with 9-electron charge resolution in room temperature, in *2018 IEEE Micro Electro Mechanical Systems (MEMS)* (IEEE, 2018), pp. 67–70
37. T.A.W. Roessig, Integrated MEMS tuning fork oscillators for sensor applications. Ph.D. thesis, UC Berkely, 1999
38. C. Zhao, M. Pandit, G. Sobreviela, A. Mustafazade, S. Du, X. Zou, A. Seshia, Supplementary: on the noise optimization of resonant MEMS sensors utilizing vibration mode localization. Appl. Phys. Lett. **112**(19), 194103 (2018)
39. C. Zhao, M. Pandit, B. Sun, G. Sobreviela, X. Zou, A. Seshia, A closed-loop readout configuration for mode-localized resonant MEMS sensors. J. Microelectromech. Syst. **26**(3), 501–503 (2017)
40. M. Pandit, C. Zhao, G. Sobreviela, A. Mustafazade, S. Du, X. Zou, A.A. Seshia, Closed-loop characterization of noise and stability in a mode-localized resonant MEMS sensor. IEEE Trans. Ultrason. Ferroelectr. Freq. Control **66**(1), 170–180 (2019)
41. C. Zhao, G.S. Wood, J. Xie, H. Chang, S.H. Pu, M. Kraft, A comparative study of output metrics for an MEMS resonant sensor consisting of three weakly coupled resonators. J. Microelectromech. Syst. **25**(4), 626–636 (2016)
42. H.K. Lee, R. Melamud, S. Chandorkar, J. Salvia, S. Yoneoka, T.W. Kenny, Stable operation of MEMS oscillators far above the critical vibration amplitude in the nonlinear regime. J. Microelectromech. Syst. **20**(6), 1228–1230 (2011)
43. L.G. Villanueva, E. Kenig, R.B. Karabalin, M.H. Matheny, R. Lifshitz, M.C. Cross, M.L. Roukes, Surpassing fundamental limits of oscillators using nonlinear resonators. Phys. Rev. Lett. **110**(17), 177208 (2013)
44. M. Pandit, C. Zhao, G. Sobreviela, S. Du, X. Zou, A. Seshia, Utilizing energy localization in weakly coupled nonlinear resonators for sensing applications. J. Microelectromech. Syst. (2019)
45. C. Zhao, G. Sobreviela, M. Pandit, S. Du, X. Zou, A. Seshia, Experimental observation of noise reduction in weakly coupled nonlinear MEMS resonators. J. Microelectromech. Syst. **26**(6), 1196–1203 (2017)
46. J. Juillard, P. Prache, N. Barniol, Analysis of mutually injection-locked oscillators for differential resonant sensing. IEEE Trans. Circ. Syst. I Regul. Pap. **63**(7), 1055–1066 (2016)
47. P. Prache, A. Uranga, N. Barniol, J. Juillard, Temperature-drift rejection and sensitivity to mismatch of synchronized strongly-coupled M/NEMS resonators, in *2016 IEEE 29th International Conference on Micro Electro Mechanical Systems (MEMS)* (IEEE, 2016), pp. 1054–1057
48. P. Prache, J. Juillard, P.M. Ferreira, N. Barniol, M. Riverola, Design and characterization of a monolithic CMOS-MEMS mutually injection-locked oscillator for differential resonant sensing. Sens. Actuators Phys. **269**, 160–170 (2018)
49. J. Juillard, P. Prache, P.M. Ferreira, N. Barniol, Ultimate limits of differential resonant MEMS sensors based on two coupled linear resonators. IEEE Trans. Ultrason. Ferroelectr. Freq. Control **65**(12), 2440–2448 (2018)
50. J. Juillard, A. Mostafa, P.M. Ferreira, Nonlinear enhancement of locking range of mutually injection-locked oscillators for resonant sensing applications, in *2018 European Frequency and Time Forum (EFTF)* (IEEE, 2018), pp. 109–113

Radio-Frequency Capacitive Gate-Based Charge Sensing for Semiconductor Quantum Dots

Imtiaz Ahmed and M. Fernando Gonzalez-Zalba

Abstract We present a comprehensible introduction to capacitive gate-based sensing, a technique for charge state readout of gate-defined semiconductor quantum dots. The objective of this chapter is to introduce the reader to the fundamental concepts necessary to understand the technique from a theoretical perspective and to be able to apply the method experimentally. We aimed to maintain a pedagogical tone with necessary rigour to keep it accessible and self-contained. We start by introducing the fundamentals of single-electron effects in single and double quantum dots i.e. Coulomb blockade. We use the constant interaction model and the effect of quantum confinement with a focus on explaining the electrical impedance of these systems under direct current excitation. We then explain how these properties are modified under the effect of an electrical excitation at high-frequency. More particularly, we focus on the appearance of a new component to the impedance, the so-called *parametric capacitance* when single-electrons tunnels back and forth between stable charge states because of the effect of the drive. We show how the *parametric capacitance* is composed by two physically distinct terms, the *quantum capacitance* and the *tunneling capacitance*, associated to adiabatic and non-adiabatic tunneling processes, respectively. In the second part of this chapter, we introduce the experimental technique. We show how embedding the devices in a high-Q lumped-element LC resonator can be used to probe the parametric capacitance of the system. We first present the fabrication technique of superconducting niobium nitride (NbN) spiral inductors and how these can be used in conjunction with the devices to construct high-Q resonators at radio-frequencies (rf). We show how the resonator can be characterized and optimized for optimum sensitivity of the charge state of the quantum dot system. Finally, we present an experimental demonstration of rf capacitive gate-based sensing utilizing a NbN inductor coupled to the gate of a silicon nanowire field effect transistor. Coulomb blockade manifest in these devices at

I. Ahmed (✉)
Department of Electrical and Electronic Engineering,
University of Dhaka, Dhaka 1000, Bangladesh
e-mail: imtiaz@du.ac.bd

M. F. Gonzalez-Zalba
Hitachi Cambridge Laboratory, J. J. Thomson Ave., Cambridge CB3 0HE, UK
e-mail: mg507@cam.ac.uk

Y. Zhu (ed.), *Micro and Nano Machined Electrometers*,
https://doi.org/10.1007/978-981-13-3247-0_3

milikelvin temperatures allowing probing their parametric capacitance when driven by an external high-frequency excitation. Based on the theoretical concepts in combination with the experiential demonstration presented, this chapter provides a pedagogical introduction to rf gated-based charge sensing.

1 Introduction

Charge dynamics dictates the electrical properties of nanoscale single electron devices. Gate-defined quantum dots (QDs) in semiconductor materials are prototypical examples of these single electron devices [1, 2]. Charge sensing in QDs can be useful for diverge range of applications in quantum metrology [3], charge noise characterization [4–6], cryogenic thermometry [7–10] and solid state quantum computation [11–14].

Rapid charge sensing with high fidelity is a crucial requirement for implementing new quantum technologies with different single electron devices. Charge sensing in QDs requires noninvasive and sensitive electrometers or charge sensors. Conventional charge sensing in QDs is performed by conductance measurements of proximally placed external charge sensors like quantum point contact (QPC) [13] or single electron transistor (SET) [15]. An alternative approach for QD charge sensing utilizes dispersive interaction between QD charge and microwave photon field. Here the QD is embedded in a high Q-factor on-chip microwave resonator where charge and microwave photon hybridize their dynamics in the strong coupling regime. In the dispersive limit, where the QD and resonator are detuned, the QD charge state can then be directly inferred from the oscillatory state of the resonator [16–20].

Based on the principle of the dispersive interaction, a compact technique for charge sensing is developed which uses existing gate electrode used to define the QD as a sensing port, hence the name gate-based sensing [21–26]. This technique uses off-chip lumped-element resonator which is coupled to the sensing port to capture the charge dynamics in the QDs using radio frequency (rf) reflectometry [21, 22, 27]. This method alleviates the burden of external electrometers like QPC and SET. Charge dynamics in the QDs gives rise to reactive changes such as the quantum or tunneling capacitance [28–31]. In this chapter we will present detail theoretical background and experimental demonstration needed to understand capacitive gate-based sensing in QDs using rf-reflectometry technique.

2 Charge Transport and Dynamics in a Single QD System

Gate-based sensors can be used to detect dispersive signals in QDs tunnel coupled to reservoirs. The dispersive signals arise due to reactive changes that occur when electrons tunnel cyclically between a QD and a reservoir causing the appearance of the tunneling capacitance [28–31]. In this section we discuss the theoretical background needed to understand the functional dependence of capacitance in single QDs.

2.1 Electron Transport Through a Single QD

2.1.1 Coulomb Blockade Transport in QD

We consider a simple single-electron tunneling system where a single QD island is weakly tunnel coupled to two electron reservoirs source (s) and drain (d), see Fig. 1a. We present a purely classical description in which the influence of discrete quantum states in the QD is neglected. The tunnel couplings between the QD and the s/d reservoirs are modelled by tunnel barriers consisting of resistors (R_s, R_d) and capacitors (C_s, C_d). The gate g is capacitively coupled to the QD through the capacitor C_g. The voltage V_g tunes the potential of the QD allowing continuous control of the electron density in the QD. The s and d electrostatic potentials are controlled by the source-drain bias voltage V_{sd}.

The current I_{sd} from the s to d terminal through the QD shows periodic oscillations as a function of the gate voltage V_g, see Fig. 1b. These oscillations are known as Coulomb oscillations. Electron transport is not allowed between the peaks due to charge quantization effect resulting from Coulomb blockade (CB); i.e., electrons can only be added to the QD from the s/d reservoirs when they have sufficient energy to overcome the Coulomb repulsion from the fixed number of electrons in the QD. Hence the name CB effect is given to characterize these oscillations which denote sequential electron transport from the s to d terminal through the QD [32, 33].

2.1.2 Conditions for Coulomb Blockade

We can relate the CB effect with different energy scales in the single QD system. The first one is the charging energy $E_c = e^2/C$ where e is the electron charge and C stands for the total capacitance as seen from the QD to the rest of the world. The charging energy E_c quantifies the minimum energy to add an extra electron to the QD. The second important energy scale is the thermal energy $k_B T$ of the system, where k_B

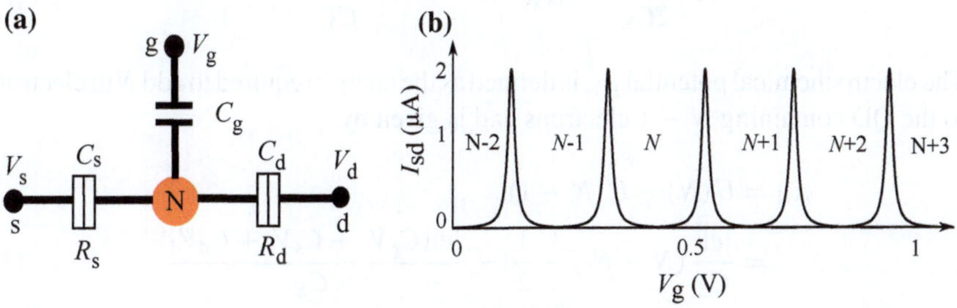

Fig. 1 **a** A QD containing N electrons is tunnel coupled (R_s, C_s and R_d, C_d) to the source (s) and drain (d) reservoirs. A capacitively coupled gate g allows electrostatic control of the QD potential through C_g. **b** Coulomb oscillations in source-drain current I_{sd} as a function of gate voltage V_g

is the Boltzmann constant and T stands for the temperature of the s/d reservoir. The first condition for the Coulomb blockade can be stated as $E_c \gg k_B T$. Typically, E_c must be ten times larger than $k_B T$ for robust operation [34]. This condition imposes some tight requirements in terms of nanofabrication and for the particular case of silicon multi-electron QDs, this sets the critical dimensions of the nanostructure below 80 nm for operation at 4 K and sub-nanometer for room temperature [35]. The second condition for CB effect dictates the tunneling resistances R_s and R_d to be greater than the quantum of resistance $R_q = h/e^2 = 25.8\ \text{k}\Omega$, where h is Planck constant. This ensures Heisenberg uncertainties in electron energies to stay below the E_c. If both conditions are met, the Coulomb energy will be dominant and the system is said to be in the CB regime.

2.1.3 The Constant Interaction Model

The CB effect can be described using the Constant Interaction (CI) model [36–39]. This model assumes that the Coulomb interaction of an electron in the QD with all other electrons, located in or outside the QD island, can be parameterized by a set of constant capacitances independent of the number of electrons N in the QD. Based on this assumption, one can write down the voltage V_0 at the QD island as

$$V_0 = \frac{Q_0 + Q_{bg} + C_g V_g + C_s V_s + C_d V_d}{C_\Sigma}, \tag{1}$$

where $C_\Sigma = C_g + C_d + C_s$, Q_{bg} and Q_0 stand for any background and initial charges on the QD respectively. The potential energy $U(N)$ of the QD with N electrons can be written as

$$\begin{aligned} U(N) &= \int_0^{-N|e|} V_0(Q_0) dQ_0 \\ &= \frac{N^2|e|^2}{2C_\Sigma} - N|e|\frac{Q_{bg} + C_g V_g + C_s V_s + C_d V_d}{C_\Sigma}. \end{aligned} \tag{2}$$

The electrochemical potential μ_N is defined as the energy required to add Nth electron to the QD containing $N-1$ electrons and is given by

$$\begin{aligned} \mu_N &= U(N) - U(N-1) \\ &= \frac{|e|^2}{C_\Sigma}(N - N_{bg} - \frac{1}{2}) - \frac{|e|(C_g V_g + C_s V_s + C_d V_d)}{C_\Sigma}. \end{aligned} \tag{3}$$

In the low bias regime $V_{sd} \approx 0$; the s/d electrochemical potentials are $\mu_s = \mu_d \approx 0$. The electrochemical potential μ_N aligns with the s/d when $V_g(N) = |e|(N - N_{bg} - \frac{1}{2})/C_g$ and electrons can tunnel through the QD island. At other values of V_g, when the electrochemical potentials are misaligned, electron transport is blocked. Therefore, a scan of the gate voltage V_g generates a series of periodic CB oscillations in Fig. 1b and the spacing between peaks is given by

$$\Delta V_g = V_g(N) - V_g(N-1) = \frac{|e|}{C_g}. \tag{4}$$

The charging energy E_c is defined as the difference in the electrochemical potentials for N and $N-1$ electrons in the QD island and is given by

$$E_c = \mu_N - \mu_{N\text{-}1} = \frac{e^2}{C_\Sigma}. \tag{5}$$

Now for non-zero V_{sd}, under symmetric bias $\mu_s = -\mu_d = -|e|V_{sd}/2$, the current I_{sd} shows characteristic Coulomb diamonds in (V_g, V_{sd}) space. This is evident in the charge stability map in Fig. 2. When $V_{sd} = 0$, for example at points like A and C, electron transport is blocked as none of the QD electrochemical potentials are aligned with μ_s and μ_d. The current I_{sd} is zero inside the diamonds due to CB effect. But at point B, electrochemical potential μ_N is aligned with μ_s and μ_d and CB is lifted that allows finite I_{sd} to flow through the QD. At finite s/d bias, electrochemical potential

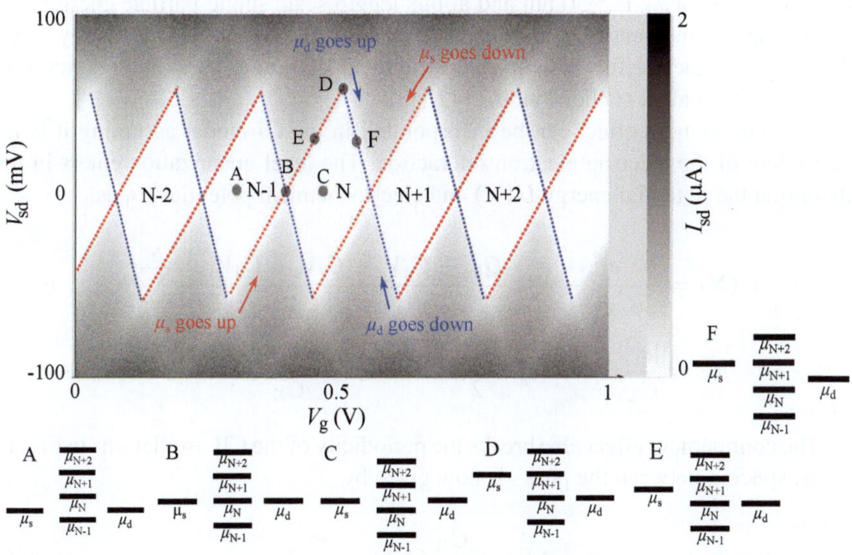

Fig. 2 Coulomb diamonds in source drain current I_{sd} in V_g and V_{sd} bias voltage space. Simulated for $C_g = C_s = C_d$ at 100 mK using SIMON software [40]. Positions of the electrochemical potentials μ for s, d and the QD are illustrated for different points A to F in the stability diagram

alignment occurs along the red and blue dashed lines. The slopes of the red (positive, +) and blue (negative, −) lines are different and controlled by the capacitances C_g, C_s and C_d as

$$\frac{V_{sd}}{V_{g}}_{+} = \frac{C_g}{C_s}, \tag{6}$$

$$\frac{V_{sd}}{V_{g}}_{-} = -\frac{C_g}{C_g + C_d}. \tag{7}$$

The charging energy E_c can be directly measured from the height of the diamonds; for example the point D in Fig. 2, where μ_s and μ_d align with μ_{N+1} and μ_N respectively. Electron transport can occur through electrochemical potentials μ_{N+1} and/or μ_N. For other $V_{sd} \neq 0$ points like E (F), the electrochemical potential μ_N aligns with μ_d (μ_N aligns with μ_s) allowing electron transport through the QD.

2.1.4 Quantum Confinement Effect

Here we discuss how quantum mechanical confinement effect modifies the classical CB model discussed in the previous section. This effect is important when the size of the QD island is comparable with the Fermi wavelength λ_F of the electron. For Si the typical values for λ_F is ∼50 nm and at this length scale single-particle energies E_n due strong confinement need to be considered [41]. The single-particle energy levels due to confinement effect resembles the energy spectrum of an atom and therefore QDs are often called artificial atoms [42, 43].

The confinement effect can be incorporated in the CI model assuming it is independent of the electron-electron interaction. The level quantization enters in the definition the potential energy $U(N)$ and electrochemical potential μ_N as

$$U(N) = \frac{N^2|e|^2}{2C_\Sigma} - N|e|\frac{Q_{bg} + C_g V_g + C_s V_s + C_d V_d}{C_\Sigma} + \sum_{n=1}^{N} E_n \tag{8}$$

$$\mu_N = \frac{|e|^2}{C_\Sigma}(N - N_{bg} - \frac{1}{2}) - \frac{|e|(C_g V_g + C_s V_s + C_d V_d)}{C_\Sigma} + E_n. \tag{9}$$

The confinement effect also breaks the periodicity of the CB oscillations in Fig. 1b as the spacing between the peaks is now given by

$$\Delta V_g = \frac{C_\Sigma}{eC_g}(\Delta E_n + \frac{e^2}{C_\Sigma}) \tag{10}$$

where $\Delta E_n = E_n - E_{n-1}$ is the energy gap between the single-particle energy levels. The energy needed to add an electron in the QD is now known as addition energy E_{add} which is given by

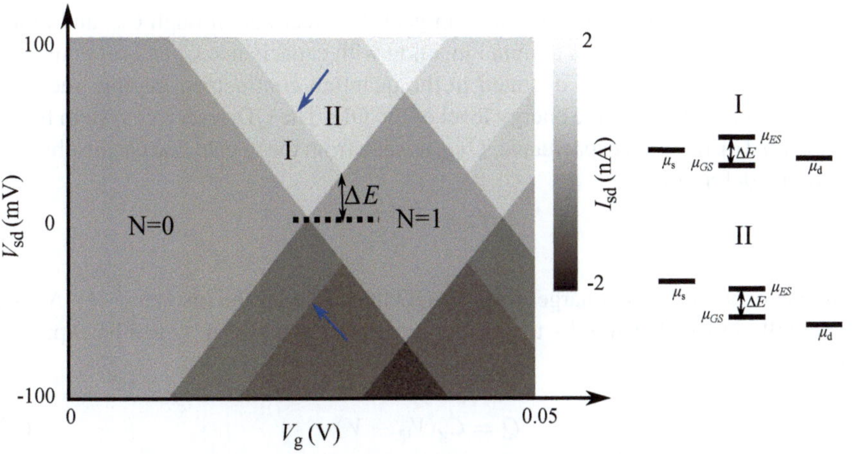

Fig. 3 Coulomb diamonds in quantum confinement regime considering the ground state (GS) and one excited state (ES). The excited state appears as parallel lines in non-zero source-drain bias marked with blue arrows. The ΔE denotes the energy separation between the GS and ES. On the right, electrochemical potentials for two different transport regions I and II in the stability map are shown. Simulation is performed for $C_g = C_s = C_d$ at 100 mK using SIMON software [40]

$$E_{\text{add}} = \frac{e^2}{C_\Sigma} + \Delta E_n. \tag{11}$$

To observe the effect of these different single-particle energy levels in the transport measurement, the thermal energy $k_B T \ll \Delta E_n$. In the single-particle energy level context, only the ground state (GS) (n = 1) energy level participates in electron transport in the previous section. Now when excited states E_n (n = 2, 3, 4, ...) are involved, they can be probed at finite source-drain bias which appear as additional features running parallel to the edges of Coulomb diamonds. We have simulated the Coulomb diamonds considering the GS and only one excited state ES, see Fig. 3. The ES of the QD for N = 1 charge state appeared as parallel lines ΔE above and below the electrochemical potential of the GS (marked by blue arrows) and these lines terminate on $N = 1$ charge state. Moreover, transport in region I only involves electrochemical potential of the GS μ_{GS} but at higher source-drain bias in region II electrochemical potential of the excited state μ_{ES} contributes along with μ_{GS}.

2.2 The Origin of Capacitance in a Single QD-reservoir System

Now, we move on to the alternating current description of a QD coupled to reservoir. To understand the origin of the capacitance in QD-reservoir systems, let us consider a QD in thermal equilibrium with an electron reservoir whose temperature is T.

The QD is capacitively coupled to a top-gate (tg) electrode through C_{tg}, and tunnel coupled to the reservoir via a tunnel junction with capacitance C_{j} and resistance R_{j}, see Fig. 4a. The system is operated in the quantum confinement regime such that electrons occupy a discrete energy level of the QD. The QD-reservoir system has an associated differential capacitance, C_{diff}, as seen from the tg gate that can be obtained from the definition

$$C_{\text{diff}} = \frac{dQ}{dV_{\text{tg}}}. \tag{12}$$

We can express the net charge Q in the QD that includes the discrete $-|e|\langle N\rangle$ and externally induced charge by the gate and reservoir electrodes. Q can be expressed as

$$Q = C_{\text{g}}(V_{\text{tg}} - V) \tag{13}$$

$$Q = C_{\text{j}}V - |e|\langle N\rangle \tag{14}$$

where V is the voltage on the QD. By eliminating V, one can write the expression for Q as

$$\begin{aligned} Q &= \frac{C_{\text{j}}C_{\text{tg}}V_{\text{tg}}}{C_{\text{j}} + C_{\text{tg}}} - \frac{C_{\text{tg}}}{C_{\text{j}} + C_{\text{tg}}}|e|\langle N\rangle \\ &= \alpha C_{\text{j}}V_{\text{tg}} - \alpha|e|\langle N\rangle \end{aligned} \tag{15}$$

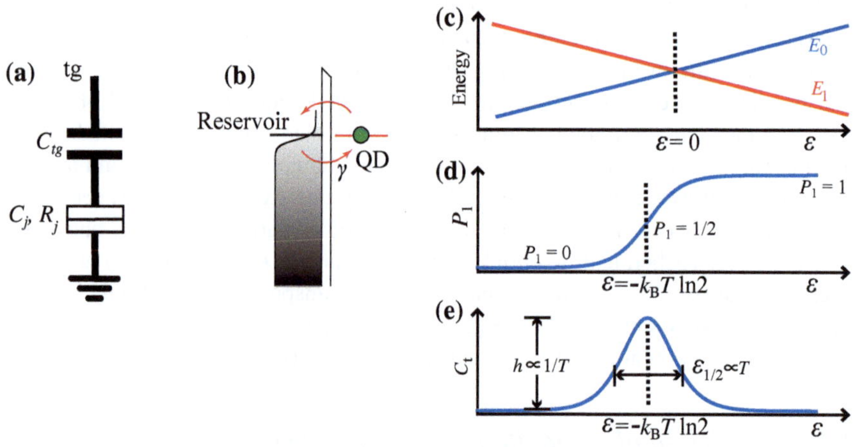

Fig. 4 **a** Equivalent circuit of the QD-reservoir system [10]. **b** Schematic of cyclic electron exchange between a discrete energy level of a QD and a thermally broadened electron reservoir [10]. **c** Energy diagram of a fast driven TLS with discrete energies E_0 and E_1, across a charge degeneracy point [10]. **d** Probability P_1 of an electron to be in the QD as a function of energy level detuning ε [10]. **e** Tunneling capacitance C_t as a function of ε [10]

where α is the gate coupling $C_{\mathrm{tg}}/(C_{\mathrm{j}}+C_{\mathrm{tg}})$. Now inserting Eq. (15) in Eq. (12) gives the differential capacitance as the sum of two components

$$C_{\mathrm{diff}}=\frac{\partial Q}{\partial V_{\mathrm{tg}}}=\underbrace{\alpha C_{\mathrm{j}}}_{\text{geometrical}}\underbrace{-|e|\alpha\frac{\partial P_1}{\partial V_{\mathrm{tg}}}}_{\text{tunneling}}=C_{\mathrm{geom}}+C_{\mathrm{t}}, \tag{16}$$

assuming $P_1\sim\langle N\rangle$ where P_1 is the probability of having an excess electron in the QD. The analytical expression for the tunneling capacitance C_{t} can be obtained considering the QD-reservoir charge distribution. In the limit of weak tunnel coupling, the QD-reservoir system can be described by the Hamiltonian $H=\frac{1}{2}\varepsilon\sigma_{\mathrm{z}}$ where ε is the energy detuning and σ_{z} is the z Pauli matrix. The eigenenergies $E_0=\varepsilon/2$ and $E_1=-\varepsilon/2$ are associated with the QD states with zero and one excess electron, respectively. This additional electron can tunnel in and out of the electron reservoir at a rate γ, as schematically depicted in Fig. 4b. The energy detuning between these states can be controlled by V_{tg} given that $\varepsilon=-e\alpha(V_{\mathrm{tg}}-V_0)$. Here V_0 is the gate voltage offset at which the two eigenstates are degenerate.

To probe the tunneling capacitance, the system is subject to a modulation occurring at some frequency, f_{c} that varies the energy detuning $\varepsilon=\varepsilon_0+\delta\varepsilon\sin(2\pi f_{\mathrm{c}}t)$. In the limit $\gamma\gg f_{\mathrm{c}}$, the QD and reservoir are in thermal equilibrium and electrons tunnel in and out of the reservoir adiabatically. In this situation, P_1 tracks the thermal population, P_1^0, given by the instantaneous gate-voltage excitation [31] and C_{t} can be expressed as

$$C_{\mathrm{t}}=-e\alpha\frac{\partial P_1^0}{\partial V_{\mathrm{tg}}}=(e\alpha)^2\frac{\partial P_1^0}{\partial\varepsilon}. \tag{17}$$

From the energy spectrum represented in Fig. 4c and taking into account the spin degeneracy of two in the QD, Maxwell-Boltzmann statistics give the equilibrium probability distribution

$$P_1^0=\frac{2\exp(\varepsilon/2k_{\mathrm{B}}T)}{\exp(-\varepsilon/2k_{\mathrm{B}}T)+2\exp(\varepsilon/2k_{\mathrm{B}}T)}, \tag{18}$$

and this is depicted as a function of detuning ε in Fig. 4d. At large negative detuning, the QD remains unoccupied ($P_1^0=0$), at large positive detuning the QD is occupied ($P_1^0=1$) and at the degeneracy point $P_1^0=1/2$. First we calculate the change in P_1^0 with respect to the energy detuning ε as

$$\frac{\partial P_1^0}{\partial\varepsilon}=\frac{1}{4k_{\mathrm{B}}T}\frac{1}{\cosh^2\left(\frac{\varepsilon}{2k_{\mathrm{B}}T}\right)}. \tag{19}$$

Using this expression in Eq. (19), one can rewrite the expression for the tunneling capacitance C_t in Eq. (17) as

$$C_t = \frac{(e\alpha)^2}{4k_B T} \frac{1}{\cosh^2(\frac{\varepsilon}{2k_B T})} \tag{20}$$

where we have redefined detuning to account for the peak-center shift induced by temperature ($\varepsilon \rightarrow \varepsilon + k_B T \ln 2$). The tunneling capacitance C_t has a full-width-half-maximum (FWHM) with respect to ε of

$$\varepsilon_{1/2} = 4 \ln(\sqrt{2} + 1) k_B T, \tag{21}$$

as plotted in Fig. 4e. This analysis is valid as long as $k_B T$ remains smaller than the discrete energy spacing in the QD (ΔE) and larger than the QD level broadening ($h\gamma$). In the latter case ($k_B T < h\gamma$), C_t takes a Lorentzian form given by

$$C_t = \frac{(\alpha e)^2}{\pi} \frac{h\gamma}{(h\gamma)^2 + \varepsilon^2}, \tag{22}$$

and $\varepsilon_{1/2}$ is given by $2h\gamma$ [44]. In the case $\gamma \leq f_c$, electrons tunnel between the QD and reservoir non-adiabatically and one can solve the master equation considering 0D QD and 3D electron reservoir to derive the expression for C_t which takes the form

$$C_t = \frac{(e\alpha)^2}{2k_B T} \frac{1}{\cosh^2(\frac{\varepsilon}{2k_B T})} \frac{(2\pi\gamma_0)^2}{(2\pi\gamma_0)^2 + (2\pi f_c)^2} \tag{23}$$

γ_0 is the constant tunnel rate that establishes away from the degeneracy [22].

The relaxation rate γ is directly linked to the shape of the tunnel barrier between the QD and the reservoir which can be tuned electrically by, for example, a gate electrode. The tunneling capacitance C_t can be probed with high-frequency techniques such as gate-based reflectometry [21, 22, 45].

3 Charge Transport and Dynamics in Double QD

Double quantum dot (DQD) system in semiconductor has gain significant attention since DiVincenzo put forward his original proposal in Ref. [46] for performing quantum computation. Both spin [13] and charge [47] qubit system have been implemented using DQD system. The DQD can be arranged in series [1, 48] or parallel configuration [22, 49, 50]. In this section we present electron transport and dynamics to explain the origin of capacitance in a DQD system.

3.1 Charge Stability Diagram for DQD System

We begin by considering electron transport through a DQD system in series configuration [48, 51, 52]. Consider the DQD structure as shown in Fig. 5a where dots QD_1 and QD_2 are in CB regime with fix number of electrons N_1 and N_2. The QD_1 (QD_2) is capacitively coupled to gate electrode $g_1(g_2)$ via $C_{g1}(C_{g2})$, tunnel coupled to left (L) and right (R) electron reservoirs through C_L and C_R respectively. The mutual capacitance between QD_1 and QD_2 is C_m. The notation (N_1, N_2) represents charge state of QD_1 with N_1 electrons and QD_2 with N_2 electrons, electrochemical potential of the L and R reservoirs are denoted by μ_L and μ_R respectively. The electron density in each QD can be tuned with gate voltages V_{g1} and V_{g2}. Now the total electrostatic energy of the DQD system with (N_1, N_2) charge configuration can be derived using CI model as

$$\begin{aligned} U(N_1, N_2) = {} & \frac{1}{2}N_1^2 E_{C1} + \frac{1}{2}N_2^2 E_{C2} + N_1 N_2 E_{C_m} - \frac{1}{|e|}C_{g1}V_{g1}(N_1 E_{C1} + N_2 E_{C_m}) \\ & - \frac{1}{|e|}C_{g2}V_{g2}(N_1 E_{C_m} + N_2 E_{C2}) \\ & + \frac{1}{|e|^2}\left[\frac{1}{2}C_{g1}^2 V_{g1}^2 E_{C1} + \frac{1}{2}C_{g2}^2 V_{g2}^2 E_{C2} + C_{g1}C_{g2}V_{g1}V_{g2}E_{C_m}\right] \end{aligned} \tag{24}$$

where $E_{C(1,2)}$ are the charging energies of $QD_{1(2)}$ and E_{Cm} is the change in the energy of one dot when an electron is added to the other dot [48].

Now for low bias condition $V_L = V_R \approx 0$, the charge stability diagram can be plotted for the current through DQDs system as a function V_{g1} and V_{g2}, see Fig. 5b.

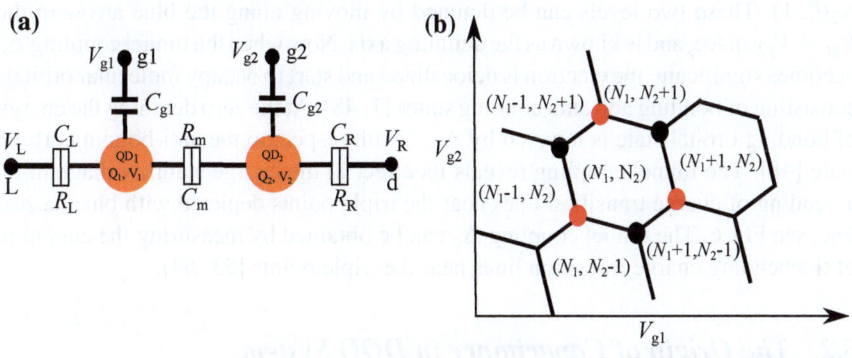

Fig. 5 **a** Double quantum dot (DQD) system. QD_1 and QD_2 are tunnel coupled (C_m, R_m) to each other. Both dots are also tunnel coupled to left (L) and the right (R) reservoirs through C_L, R_L and C_R, R_R respectively. QDs potentials are controlled by capacitively coupled gates g_1 and g_2 through C_{g1} and C_{g2}. **b** Characteristic honeycomb pattern in charge stability diagram for a DQD system. The solid black lines correspond to charge transition in the DQD system. Any two charge states sharing the charge transition lines as a common boundary, will be degenerate

The charge stable regions with fixed (N_1, N_2) form characteristic hexagonal honeycomb pattern and charge instability occurs at the boundaries between two adjacent hexagons. The six boundaries of the (N_1, N_2) can be found equating total energies $U(N_1, N_2)$ of the neighbouring stable charge state. For example the boundary line between (N_1, N_2) and $(N_1 + 1, N_2)$ occurs when $U(N_1, N_2) = U(N_1 + 1, N_2)$; i.e., the two charge states are degenerate in energy and electrons are exchanged between L reservoir and QD_1. The electrochemical potential $\mu_{1(2)}(N_1, N_2)$ of $QD_{1(2)}$ is defined as the energy required to add $N_{1(2)}$th electron to $QD_{1(2)}$, while having $N_{2(1)}$ electrons in $QD_{2(1)}$. The expression for these electrochemical potentials are given by

$$\mu_1(N_1, N_2) = (N_1 - \frac{1}{2})E_{C1} + N_2 E_{Cm} - \frac{1}{|e|}(C_{g1} V_{g1} E_{C1} + C_{g2} V_{g2} E_{Cm}), \quad (25)$$

$$\mu_2(N_1, N_2) = (N_2 - \frac{1}{2})E_{C2} + N_1 E_{Cm} - \frac{1}{|e|}(C_{g1} V_{g1} E_{Cm} + C_{g2} V_{g2} E_{C2}). \quad (26)$$

3.1.1 One Electron DQD system

The charge stability diagram for a general $(N_1 + 1, N_2)$–$(N_1, N_2 + 1)$ transition in a few electron DQDs system is simulated in Fig. 6. This picture can be used to describe a DQD system in one electron regime by considering $N_1 = 1$ and $N_2 = 0$, i.e., (1,0)-(0,1) charge configuration in the charge stability diagram. The charge transition lines A, B, C and D denote electron transport between QDs and electron reservoirs (L, R). For example, the line A denotes the alignment of electrochemical potential $\mu_1(1, 0)$ with μ_L so that electron exchange takes place between QD_1 and L reservoir. The line E represents the inter-dot transition (ICT) where electron tunnels between the two QDs. This happens when the electrochemical potential $\mu_1(1, 0)$ aligns with $\mu_2(0, 1)$. These two levels can be detuned by moving along the blue arrow in the $V_{g1} - V_{g2}$ space, and is known as the detuning axis. Now when the tunnel coupling Δ_c becomes significant, the electron is delocalized and start to occupy molecular orbitals consisting of bonding and anti-bonding states [1, 48]. At the zero detuning the energy of bonding orbital state is lowered by $\Delta_c/2$ with respect to the anti-bonding orbital state [48]. The tunnel coupling reveals its effect in the charge stability diagram by a bending of charge transition lines near the triple points depicted with blue dashed line, see Fig. 6. The tunnel coupling Δ_c can be obtained by measuring the curvature of the bending charge transition lines near the triple points [53, 54].

3.2 *The Origin of Capacitance in DQD System*

3.2.1 DQD Capacitance Model

Now, we move on to the alternating current (AC) description of a DQD. To illustrate the origin of AC capacitance in double quantum dot systems, we consider a DQD system, see Fig. 7a, in which the two quantum dots (QDs) 1 and 2 are coupled to a

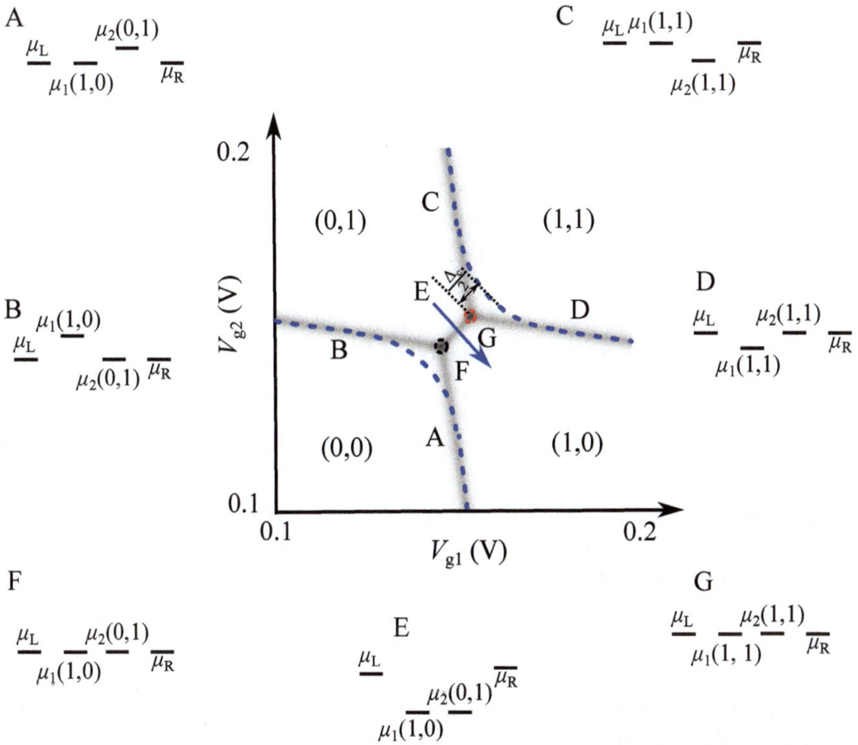

Fig. 6 Charge stability diagram for a DQD system at (1,0)–(0,1) charge transition region, simulated for $C_L = C_R = C_L = 2C_m$ using SIMON software [40]. Electrochemical potentials μ for L, R, QD_1 and QD_2 are shown for different regions in the stability diagram. Four different charge states, (0,0), (1,0), (0,1) and (1,1) can be distinguished. The triple points are marked by black (G) and red (F) circles. At the charge transition line (marked by E) connecting the two triple points, the charge states (0,1) and (1,0) are degenerate. The blue arrow marks the detuning axis for the DQD. At the other four lines A, B, C and D the electrochemical potential of at least one dot is zero and thus equals the electrochemical potential of the reservoirs L and R. The effect of tunnel coupling Δ_c is depicted in blue dashed line

sensing gate electrode (g) via gate capacitances C_{gi} for $i = 1, 2$. Each QD is further coupled to left and right electrodes by capacitances C_{Li} and C_{Ri} respectively and they are further coupled to each other by a mutual capacitance C_m. Here, we study the zero-bias regime in which left and right electrodes are connected to ground.

The DQD has an associated differential capacitance, C_{diff}, as seen from the gate that can be obtained from the following expression

$$C_{diff} = \frac{d(Q_1 + Q_2)}{dV_g}. \tag{27}$$

We can express the net charges Q_i in QD i by considering the total existing charge in the islands $-e\langle N_i\rangle$, and the polarization charge induced by the gate electrode and the mutual electrostatic effect between dots

$$\begin{aligned} Q_1 &= C_{g1}(V_G - V_1) \\ Q_1 &= -e\langle N_1\rangle + (C_{L1} + C_{R1})V_1 - C_m(V_2 - V_1) \\ Q_2 &= C_{g2}(V_g - V_2) \\ Q_2 &= -e\langle N_2\rangle + (C_{L2} + C_{R2})V_2 - C_m(V_1 - V_2). \end{aligned} \tag{28}$$

Here V_i and $\langle N_i\rangle$ are the voltage and average electron number in QD i. Solving Eqs. (28) for $Q_1 + Q_2$ through elimination of V_1 and V_2, we arrive at the simplified expression

$$Q_1 + Q_2 = e(\alpha_1\langle N_1\rangle + \alpha_2\langle N_2\rangle)[\alpha_1(C_{L1} + C_{L2}) + \alpha_2(C_{R1} + C_{R2})]V_g \tag{29}$$

in the limit of $C_m \ll C_{\Sigma i}$. Here $C_{\Sigma i} = C_{Li} + C_{Ri} + C_{gi} + C_m$ is the total capacitance of QD i and α_i its gate coupling defined as $C_{gi}/C_{\Sigma i}$. Inserting Eq. (29) into Eq. (27) gives the differential capacitance as the sum of two components

$$C_{diff} = C_{geom} + C_{pa}. \tag{30}$$

The geometrical capacitance, C_{geom}, a voltage-independent term, represents the DC limit of the capacitance

$$C_{geom} = \alpha_1(C_{L1} + C_{R1}) + \alpha_2(C_{L2} + C_{R2}). \tag{31}$$

The second term in Eq. (30), C_{pa}, represents the parametric dependence of the time-averaged excess electron numbers on the gate voltage, and is thus coined the parametric capacitance. This is written as

$$C_{pa} = -e\left[\alpha_2\frac{\partial\langle N_2\rangle}{\partial V_g} + \alpha_1\frac{\partial\langle N_1\rangle}{\partial V_g}\right] = -e\alpha'\frac{\partial\langle N_2\rangle}{\partial V_g} \tag{32}$$

since $\frac{\partial\langle N_1\rangle}{\partial V_g} = -\frac{\partial\langle N_2\rangle}{\partial V_g}$ when QDs exchange electrons and $\alpha' = (\alpha_2 - \alpha_1)$.

3.2.2 High Frequency DQD Model

The equivalent circuit of the DQD at finite frequencies is depicted inside the dashed rectangle of Fig. 7b where the geometrical and parametric capacitance appear in parallel between electrode g and ground. The simplicity of this equivalent circuit can be exploited to obtain information of the charge and spin state of the DQD. We focus on the situation in which the state of the DQD is probed dispersively at finite-

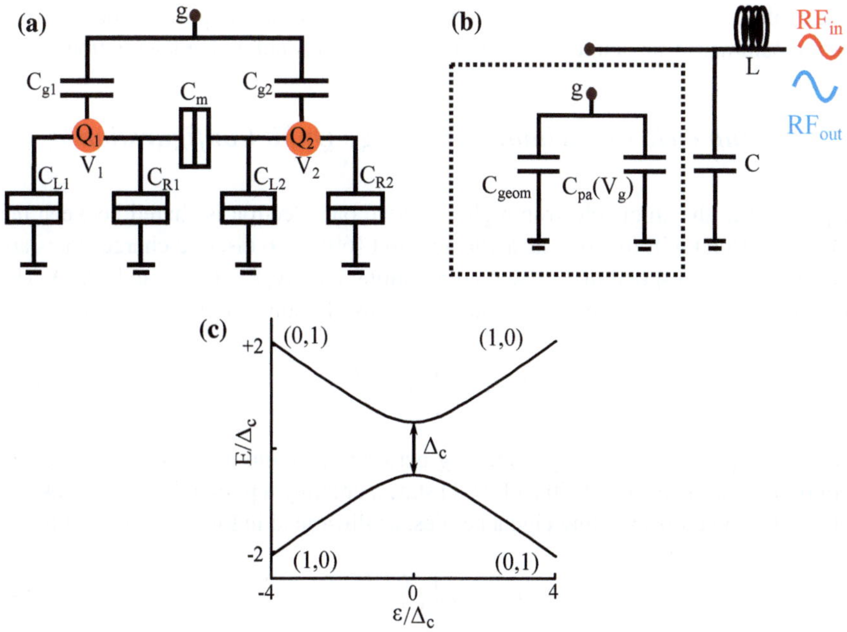

Fig. 7 **a** A schematic of a DQD system composed of QDs 1 and 2 coupled to a gate electrode g by capacitances C_{g1} and C_{g2}, respectively [31]. **b** Equivalent circuit of (**a**) including the geometrical and parametric capacitance in parallel. The resonator, with characteristic inductance L and capacitance to ground C creates an oscillatory voltage on the gate of the DQD [31]. **c** Energy level spectrum of a DQD as a function of detuning in the single electron regime. Mixing of the states forms a finite gap of magnitude Δ_c at $\varepsilon = 0$ [31]

frequencies by monitoring the state of a classical resonator coupled to the system via the gate electrode g, as shown in Fig. 7b. Here, we depict a lumped-element resonator composed by an inductance L and a capacitance to ground C and hence with a natural angular frequency of resonance given by $\omega_{res} = 1/\sqrt{LC}$. We treat the DQD-resonator system semiclassically: a quantum element coupled to a classical resonator ($\hbar\omega_c \ll k_B T, \Delta_c$). When the resonator is probed at a finite frequency ω_{res}, it produces an oscillatory gate voltage of amplitude δV_g given by

$$V_g(t) = \delta V_g \sin(\omega_{res} t), \tag{33}$$

which in turn produces a redistribution of charges Q_1 and Q_2 in the DQD due to the associated differential capacitance in Eq. (30). This additional state-dependent capacitance loads the resonator leading to a dispersive shift of the resonant frequency. This frequency shift can be readily monitored using standard homodyne detection techniques, such as reflectometry [21, 22, 27, 55–57] and transmission [58, 59]. In the following sections, we present a description of the DQD-resonator interaction.

We explore the DQD in the charge qubit configurations and exploit the dependence of the parametric capacitance on external variables for qubit characterization.

3.3 Parametric Capacitance in Charge Qubit Configuration

Operating in the single electron regime, where one electron is shared between the QDs, the DQD will function as a charge qubit [60]. Two discrete charge states are considered, corresponding to the configurations $(N_1, N_2) = (1, 0)$ and $(0, 1)$. The Hamiltonian of such a two-level system in terms of Pauli spin matrices $\sigma_{x,y,z}$ is

$$H = -\frac{\Delta_c}{2}\sigma_x - \frac{\varepsilon}{2}\sigma_z, \tag{34}$$

where ε represents the energy detuning between quantum dots and Δ_c the tunnel coupling that mixes the (1, 0) and (0, 1) states, creating a point of avoided crossing at $\varepsilon = 0$. The corresponding eigenenergies, as illustrated in Fig. 7c, are given by

$$E_{\pm} = \pm\frac{1}{2}\sqrt{\varepsilon^2 + \Delta_c^2}. \tag{35}$$

We can calculate the parametric capacitance of a charge qubit by taking into account that an oscillatory voltage on the gate will induce an oscillatory variation on the DQD energy detuning centered around ε_0 as $\varepsilon(t) = \varepsilon_0 + \delta\varepsilon \sin(\omega_{\text{res}} t)$, where $\delta\varepsilon = -e\alpha^{'}\delta V_g$. Then Eq. (32) becomes

$$C_{pa} = (e\alpha^{'})^2 \frac{\partial \langle N_2 \rangle}{\partial \varepsilon} \tag{36}$$

and the problem simplifies to calculating $\langle N_2 \rangle (\varepsilon)$.

3.3.1 The Expression for C_{pa}

In the energy basis, the time-averaged excess electron number in QD_2 in the single electron regime may be expressed as

$$\langle N_2 \rangle = \langle N_2 \rangle_- P_- + \langle N_2 \rangle_+ P_+, \tag{37}$$

where $P_{\pm}$ are the probabilities of the excess electron occupying the ground (−) or excited (+) state and

$$\langle N_2 \rangle_{\pm} = \frac{1}{2}\left(1 \pm \frac{\varepsilon}{\Delta E}\right) \tag{38}$$

are the average number of electrons in QD_2 in the ground and excite state [61]. Here, $\Delta E = E_+ - E_- = \sqrt{\varepsilon^2 + \Delta_c^2}$. Using Eqs. (36)–(38), we can then obtain the instantaneous parametric capacitance of a charge qubit:

$$C_{\mathrm{pa}}(t) = C_0\Bigg(\underbrace{\frac{\Delta_{\mathrm{c}}^3}{\Delta E^3}\chi_{\mathrm{ch}}}_{\text{quantum}} + \underbrace{\frac{\varepsilon\Delta_{\mathrm{c}}}{\Delta E}\frac{\partial\chi_{\mathrm{ch}}}{\partial t}\frac{\partial t}{\partial\varepsilon}}_{\text{tunneling}}\Bigg) \tag{39}$$

where $C_0 = (e\alpha')^2/2\Delta_{\mathrm{c}}$, $\chi_{\mathrm{ch}} = P_- - P_+$, the difference between ground and excited state occupation probabilities and we have expanded $\frac{\partial\chi_{\mathrm{ch}}}{\partial\varepsilon} = \frac{\partial\chi_{\mathrm{ch}}}{\partial t}\frac{\partial t}{\partial\varepsilon}$. Equation (39) contains two detuning-dependent contributions. The first term, corresponds to the so-called quantum capacitance arising from adiabatic transitions in systems with finite curvature of the energy bands [62, 63]. The second term, the tunneling capacitance, appears when non-adiabatic processes, such as relaxation and thermal or resonant excitation, occur at a rate comparable or faster than the probing frequency ω_{c} [22, 28, 29, 61].

Typically, in reflectometry measurements, $C_{\mathrm{pa}}(t)$ is averaged over a resonator cycle. Hence, calculating the averaged parametric capacitance $\langle C_{\mathrm{pa}}\rangle$ requires a time-dependent solution of the probability difference χ_{c}. This can be done by analyzing the dynamics of the driven two-level system which is given by the following master equation

$$\begin{aligned}\dot{P}_- &= \gamma_- P_+ - \gamma_+ P_- \\ \dot{P}_+ &= -\gamma_- P_+ + \gamma_+ P_-,\end{aligned} \tag{40}$$

where $\gamma_- = \gamma_1^{\mathrm{ch}}(1 + n_{\mathrm{p}})$ is the relaxation rate from the excited state to the ground state, $\gamma_+ = \gamma_1^{\mathrm{ch}} n_{\mathrm{p}}$ is the relaxation rate from the ground state to the excited state, γ_1^{ch} is the charge relaxation rate and $n_{\mathrm{p}} = (e^{\Delta E/k_{\mathrm{B}}T} - 1)^{-1}$ is the phonon occupation number [64, 65]. We note that γ_1^{ch} may have a functional dependence on ΔE. Such dependency, that varies for different physical implementations of the charge qubit [66–68], will not be be treated here.

We solve Eq. 40, in the limit $\hbar\gamma_1^{\mathrm{ch}} \ll k_{\mathrm{B}}T$, to first order approximation with respect to small variations of detuning $\delta\varepsilon \ll \Delta_c$ [22, 30]. In this approximation, the probability difference can be expressed as $\chi_{\mathrm{ch}}(t) = \chi_{\mathrm{ch}}^0 + \delta\chi_{\mathrm{ch}}(t)$ where $\delta\chi_{\mathrm{ch}}(t)$ represents a small deviation from the equilibrium occupation probability difference

$$\begin{aligned}\chi_{\mathrm{ch}}^0 &= \frac{1}{Z_{\mathrm{ch}}^0}\left(e^{\Delta E(\varepsilon_0)/2k_{\mathrm{B}}T} - e^{-\Delta E(\varepsilon_0)/2k_{\mathrm{B}}T}\right) \\ &= \tanh(\Delta E(\varepsilon_0)/2k_{\mathrm{B}}T),\end{aligned} \tag{41}$$

where $Z_{\mathrm{ch}}^0 = \exp[\Delta E(\varepsilon_0)/2k_{\mathrm{B}}T] + \exp[-\Delta E(\varepsilon_0)/2k_{\mathrm{B}}T]$ is the partition function for the charge qubit. We solve Eq. (40) for $\delta\chi_{\mathrm{ch}}(t)$ and obtain the following expression

$$\delta\chi_{\mathrm{ch}}(t) = -\frac{2\eta_{\mathrm{ch}}}{\omega_{\mathrm{res}}^2 + \gamma_{\mathrm{ch}}^2}\left[\gamma_{\mathrm{ch}}\sin(\omega_{\mathrm{res}}t) - \omega_{\mathrm{res}}\cos(\omega_{\mathrm{res}}t)\right]\delta\varepsilon \tag{42}$$

where η_{ch} and γ_{ch} are temperature-dependent coefficients given by

$$\eta_{ch} = -\gamma_1^{ch} \frac{1}{2k_B T} \frac{\varepsilon_0}{\Delta E(\varepsilon_0)} \frac{1}{\cosh\left(\frac{\Delta E(\varepsilon_0)}{2k_B T}\right) \sinh\left(\frac{\Delta E(\varepsilon_0)}{2k_B T}\right)}$$
$$\gamma_{ch} = \gamma_1^{ch} \coth(\Delta E(\varepsilon_0)/2k_B T). \tag{43}$$

Calculating the instantaneous parametric capacitance to first order approximation in $\delta\varepsilon$ using Eqs. (41)–(43) and averaging over a resonator cycle we obtain an analytical expression for parametric capacitance of a charge qubit

$$\langle C_{pa} \rangle \approx C_0 \left\{ \frac{\Delta_c^3}{\Delta E^3(\varepsilon_0)} \chi_{ch}^0 + \frac{\Delta_c}{2k_B T} \frac{\varepsilon_0^2}{\Delta E(\varepsilon_0)^2} \sinh^{-2}\left(\frac{\Delta E(\varepsilon_0)}{2k_B T}\right) \frac{\gamma_1^{ch2}}{\omega_{res}^2 + \coth^2\left(\frac{\Delta E(\varepsilon_0)}{2k_B T}\right) \gamma_1^{ch2}} \right\} \tag{44}$$

3.3.2 Slow and Fast Relaxation Limits

In the limit, $\gamma_1^{ch} \ll \omega_{res}$, in which the charge relaxation time is longer than the period of the resonator, thermal repopulation does not occur on the time-scale of the resonator and hence $\partial \chi_{ch}/\partial t = 0$, represented schematically on the left side of Fig. 8a. In such cases, the parametric capacitance will be composed solely of quantum capacitance contributions

$$\langle C_{pa} \rangle \approx C_0 \frac{\Delta_{ch}^3}{\Delta E^3(\varepsilon_0)} \chi_{ch}^0. \tag{45}$$

Conversely, for $\gamma_1^{ch} \gg \omega_{res}$, the thermal population distribution tracks the instantaneous excitation from the resonator ($\partial \chi_{ch}/\partial \varepsilon = \partial \chi_{ch}^0/\partial \varepsilon_0$) as schematically depicted in the right side of Fig. 8a and therefore both quantum and tunneling capacitances must be considered

$$\langle C_{pa} \rangle \approx C_0 \left\{ \frac{\Delta_{ch}^3}{\Delta E^3(\varepsilon_0)} \chi_{ch}^0 + \frac{\varepsilon_0 \Delta_{ch}}{\Delta E(\varepsilon_0)} \frac{\partial \chi_{ch}^0}{\partial \varepsilon_0} \right\}. \tag{46}$$

In Fig. 8b–d, we present a lineshape analysis of the parametric capacitance of a charge qubit as a function of temperature. The slow (Fig. 8c) and fast-relaxation regimes (Fig. 8d) exhibit a symmetric capacitance lineshape with a maximum at the avoided crossing point ($\varepsilon_0 = 0$). Here, we plot the lineshape for increasing temperature from $k_B T/\Delta_c = 0.1$ (red line) to $k_B T/\Delta_c = 5$ (light blue line). The maximum parametric capacitance diminishes with increasing temperature according to $\tanh(\Delta_c/2k_B T)$ as can be seen in both insets and it becomes vanishingly small for values $k_B T/\Delta_c > 10$. This is found to be identical under both regimes due to the tunneling capacitance having no contribution at the avoided crossing. Moreover, in the low temperature limit both regimes show a full width at half maximum, FWHM $= 1.53\Delta_c$, see in Fig. 8b. However, a distinction between the two regimes may be made through contrasting the temperature dependence of the FWHM for $k_B T/\Delta_c > 0.2$. In the slow-relaxation limit the FWHM increases until it saturates

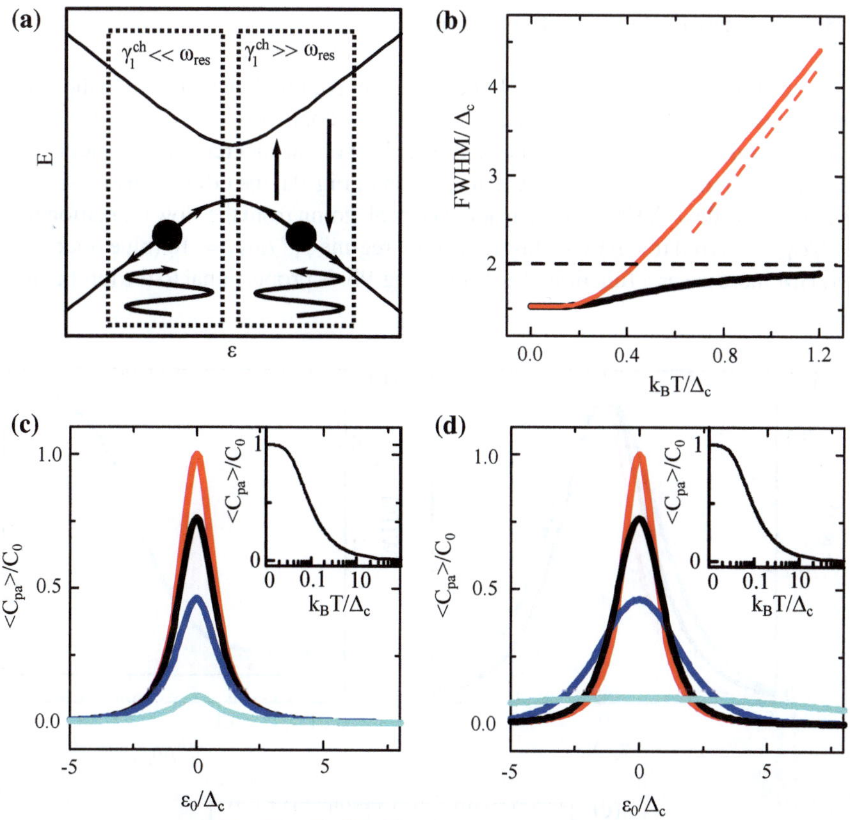

Fig. 8 Charge Qubit [31]. **a** Schematic energy level diagram of the DQD indicating the electron dynamics in the slow ($\gamma_1^{ch} \ll \omega_{res}$) and fast ($\gamma_1^{ch} \gg \omega_{res}$) relaxation regime. Only in the fast relaxation regime the system can track the instantaneous thermal population of the system exemplified by the vertical black arrows. **b** FWHM of the parametric capacitance lineshape in the slow (black solid line) and fast relaxation regime (red solid line). The slow regime tends asymptotically to FWHM= $2\Delta_c$ (black dashed line) and the fast regime to FWHM= $3.53k_BT$ (red dahsed line). Parametric capacitance lineshape as a function of detuning for $k_BT/\Delta_c = 0.1, 0.5, 1, 2$ (red, black, dark blue, light blue solid lines) in the slow (**c**) and fast (**d**) relaxation regimes. Insets indicate the maximum parametric capacitance as a function of increasing temperature for both regimes

to a value $2\Delta_c$ at high temperatures. Conversely, in the fast-relaxation regime, the FHWM shows an increase with temperature due to the tunneling capacitance. This increase tends asymptotically to a linear dependence with T at high temperatures, FWHM = $3.53k_BT$ (red dashed line). This different behaviour of the FWHM as a function of temperature allows identifying the relaxation regime in which the charge qubit is. Moreover, at low temperatures, a fit to the FWHM allows measuring Δ_c independently of the relaxation regime in which the qubit is. Finally, the fast relaxation regime at high temperatures allows calibrating the electron temperature of the system.

3.3.3 Intermediate Relaxation Regime

In the limit, $\gamma_1^{ch} \approx \omega_{res}$, in which the charge relaxation time is similar to the period of the resonator, the tunneling capacitance term contributes partially and the full Eq. (44) must be used. This regime has been observed in superconducting qubits [30] and silicon quantum dots [22]. As can be seen in Fig. 9a, the effect of increasing γ_1^{ch} is to increase the FWHM of the capacitance peak going from the slow relaxation limit $\gamma_1^{ch}/\omega_{res} = 0$ (red trace) to the fast relaxation regime $\gamma_1^{ch}/\omega_{res} = 10$ (blue trace). The FWHM increase as a function of $\gamma_1^{ch}/\omega_{res}$, Fig. 9b, is proportional to the temperature

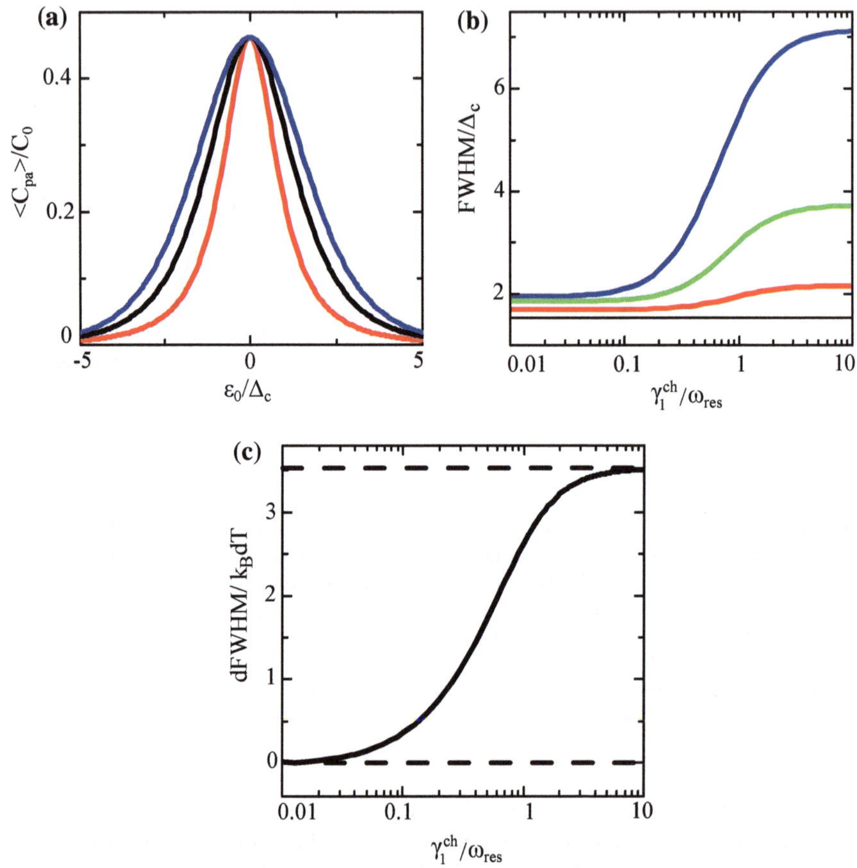

Fig. 9 Charge qubit relaxation [31]. **a** Parametric capacitance lineshape as a function of detuning for $\gamma_1^{ch}/\omega_{res} = 0$, 1 and 10 (red, black, blue lines). **b** FHWM of the parametric capacitance peak as a function of reduced charge relaxation rate $\gamma_1^{ch}/\omega_{res}$ for $k_B T/\Delta_c = 0.1, 0.5, 1$ and 2 (black, red, green and blue lines respectively). **c** Change in FWHM with respect to temperature as a function of reduced charge relaxation rate $\gamma_1^{ch}/\omega_{res}$. The dashed horizontal lines at 0 and 3.53 indicate the rate of change in the slow and fast relaxation limits respectively

of the system, going from no change at low temperatures (black line at $k_BT/\Delta_c = 0.1$) to FWHM $\approx 3.53k_BT$ at high temperatures (blue line at $k_BT/\Delta_c = 2$). Finally, in Fig. 9c, we explore the possibility of determining γ_1^{ch} from a measurement of the FWHM vs T. As can be seen in Fig. 9c, the slope dFWHM/dT, varies from 0 at the slow relaxation limit to $3.53k_B$ in the fast relaxation regime. Hence a fit to the linear region of the dependence of FWHM with T allows a direct measurement of $\gamma_1^{ch}/\omega_{res}$.

4 Capacitive Gate-Based Sensing: Theory

The capacitance in the single or double QD systems is measured dispersively at high frequencies by monitoring the state of a classical resonator coupled to the quantum system via the gate electrode g, hence the name gate-based capacitive sensing. The high frequency resonator is composed of an inductor L and a capacitor to ground C giving rise to a natural frequency of resonance which is given by

$$f_{res} = \frac{\omega_{res}}{2\pi} = \frac{1}{2\pi\sqrt{LC}}. \tag{47}$$

To probe the coupled resonator-quantum device system, rf signal RF_{in} at the resonant frequency f_{res} is sent via the gate g. This produces an oscillatory gate voltage of amplitude δV_g given by

$$V_g(t) = \delta V_g \sin(2\pi f_{res} t), \tag{48}$$

which in turn produces a redistribution of charges in the quantum element due to the associated differential capacitance, see for example in Eq. (30). This additional state-dependent capacitance loads the resonator leading to a dispersive shift of the resonant frequency f_{res}. This frequency shift can be readily monitored by detecting the reflected signal RF_{out} using standard homodyne detection techniques, such as reflectometry, hence the name rf reflectometry [21, 22, 27, 55, 56]. A more detail description of rf reflectometry is presented in Sect. 6.

The sensitivity of the gate-based sensing technique relies on the optimization of the high frequency resonator. In this section we present high-frequency resonator construction, characterization and optimization techniques.

4.1 High Frequency Resonator Construction

The high frequency resonator can be built by an inductor L, resistor R_d and a variable capacitor C_0 in parallel configuration, see Fig. 10a. The resistor R_d represents losses in the resonator; and the inductor L and the capacitance C_0 specify the resonant frequency f_{res} of the resonator. The resonator is coupled to the external rf feed line which has an impedance $Z_0 = 50\ \Omega$. The coupling between the resonator and the external circuitry is controlled by a coupling capacitor C_c.

4.2 High Frequency Resonator Characterization

There are four important parameters that characterize the resonator; (1) the resonant frequency f_{res}, (2) the full-width-half maxima (FWHM), also known as the bandwidth (BW) of the resonator, (3) loaded Q-factor $Q_L = f_{res}/\mathrm{BW}$, and (4) the nature of coupling to the external line; commonly specified by coupling constant g. All of these parameters can be measured by exciting the resonator with an external rf sinusoidal carrier with frequency f_{res} and monitoring the reflected signal. The physical quantity of interest is the complex reflection coefficient Γ which is defined as

$$\Gamma = \frac{Z - Z_0}{Z + Z_0}, \tag{49}$$

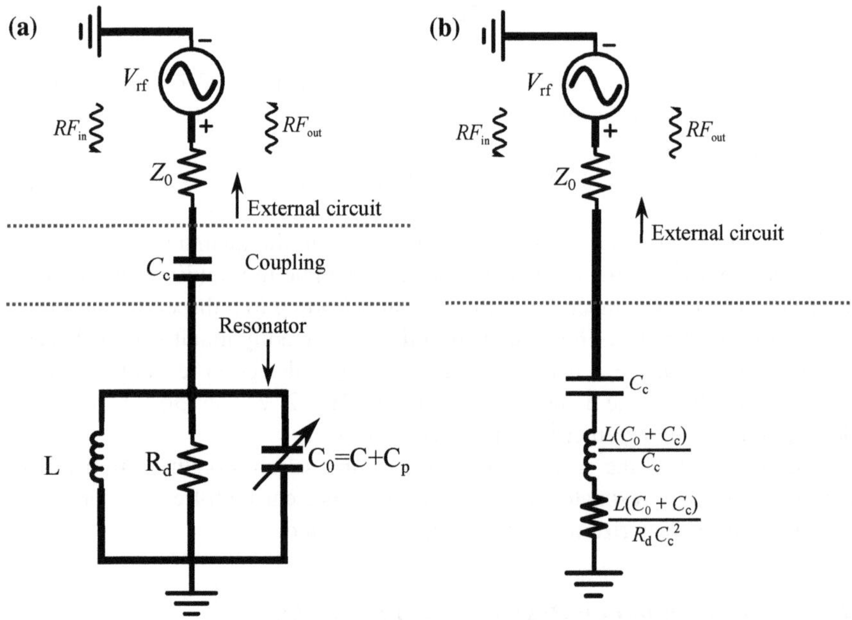

Fig. 10 **a** A high frequency resonator by connecting an inductor L, resistor R_d and a variable capacitor $C_0 = C + C_p$ in parallel configuration. The C_p is the parasitic capacitance of the resonator. The $R_d L C_0$ resonator is coupled to line impedance Z_0 through a coupling capacitor C_c. **b** A simpler equivalent circuit for the resonator at resonant frequency

where Z is the complex impedance of the coupling capacitor C_c in series with the parallel combination of the inductor L, the circuit resistance R_d and circuit variable capacitance C_0, see Fig. 10a. One can write explicitly this expression for Z as:

$$Z = R_d \frac{i\omega\Delta\omega_0}{\omega_0^2 - \omega^2 + i\omega\Delta\omega_0} + \frac{1}{i\omega C_c} \tag{50}$$

where ω is the angular frequency in rad/s, $\Delta\omega_0 = 1/(R_d C_0)$ is the line-width and $\omega_0 = 2\pi f_0 = 1/\sqrt{LC_0}$ is the unloaded angular resonant frequency of the parallel LR_dC_0 circuit. Now using the expression for Z from Eq. (50), the expression for Γ in Eq. (49) can be written as

$$\Gamma = 1 - \frac{i2g}{1+g}\left[\frac{\omega\Delta\omega}{\omega_L^2 - \omega^2 + i\omega\Delta\omega}\right] \tag{51}$$

where $\Delta\omega$ denotes the loaded linewidth, $\omega_L/2\pi = f_{res}$ is the loaded resonant frequency and g is the coupling constant and can be expressed as

$$\Delta\omega = \frac{\alpha_i + Z_0 C_c(\omega_0^2 - \omega^2)}{1 + \alpha_i C_c(R_d + Z_0)}, \; [\alpha_i = \frac{1}{R_d C_0}] \tag{52}$$

$$f_{res} = \frac{\omega_L}{2\pi} = \frac{1}{2\pi\sqrt{L(C_c + C_0)}}, \tag{53}$$

$$g = Z_0 C_c^2 \omega^2 (R_d + Z_0). \tag{54}$$

There are three Q-factors that characterize the resonator; the unloaded Q-factor Q_0, the external Q-factor Q_e and the loaded Q-factor Q_L which can be expressed as

$$Q_0 = \omega_L R_d (C_c + C_0) \tag{55}$$

$$Q_e = \frac{C_c + C_0}{\omega_L Z_0 C_c^2} \tag{56}$$

$$\frac{1}{Q_L} = \frac{1}{Q_0} + \frac{1}{Q_e}. \tag{57}$$

By comparing Eqs. (54)–(56), it is evident that the coupling constant $g = Q_0/Q_e$ and the Eq. (57) can be rewritten as

$$Q_L = \frac{Q_0}{1+g}. \tag{58}$$

The reflection coefficient $\Gamma = |\Gamma| \exp(i\varphi)$ is defined by its magnitude $|\Gamma|$ and phase $\phi = \arg(\Gamma)$. When the rf carrier frequency f_c matches with the resonant frequency f_{res}, the resonator absorbs maximum energy and the magnitude $|\Gamma|$ is at its minimum $|\Gamma|_{min}$, see Fig. 11a. The $|\Gamma|_{min} = 0$ at $f_c = f_{res}$ indicates that the resonator is critically coupled to the external line impedance Z_0. This implies perfect matching ($g = 1$); which means equal amount of power being dissipated in the resonator and the external circuitry. Away from the resonant frequency most of the power is reflected and $|\Gamma| \rightarrow 1$. The BW can be obtained directly from the FWHM of the $|\Gamma|$ vs. f_c curve. This BW represents the losses in the resonator; effectively defining the loaded Q-factor $Q_L = f_{res}/\text{BW}$. Now as the coupling deviates from its critical value $g = 1$, $|\Gamma|_{min}$ starts to increase, see Fig. 11b. Simulation shows that at resonance, $|\Gamma|_{min}$ increases for $g < 1$ or $g > 1$, see Fig. 12a. This makes the nature of coupling ambiguous from magnitude $|\Gamma|$ measurements. But the phase φ specifies whether the resonator is overcoupled $g > 1$ or undercoupled $g < 1$, see Fig. 12b. For undercoupled resonator, the phase φ traverses from negative to positive values across the resonant frequency; whereas for overcoupled resonator phase φ goes from $+180°$ to $-180°$. Hence the resonator can be fully characterized from reflection coefficient Γ measurements.

4.3 *Resonator Optimization for Reflectometry*

Now gate-based reflectometry is sensitive to changes in the reflection coefficient Γ due to the changes in capacitance C_0. Therefore, the absolute value of the differential change in reflection coefficient with C_0, i.e., $|\Delta\Gamma| = \left|\frac{\partial\Gamma}{\partial C_0}\Delta C_0\right|$ needs to be calculated. This can be done by differentiating Γ in Eq. (51) with respect to the loaded resonant frequency ω_L to find

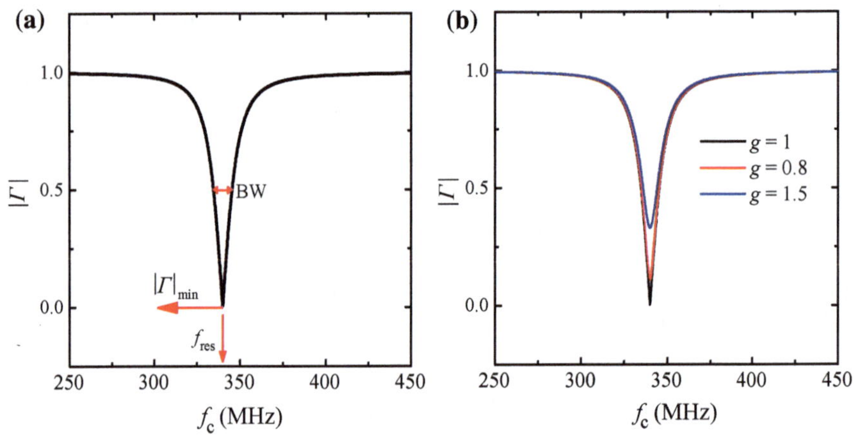

Fig. 11 **a** The magnitude $|\Gamma|$ as a function of rf carrier frequency f_c. **b** The magnitude $|\Gamma|$ as a function of rf carrier frequency f_c for different values of coupling constant g. Simulation parameters are $C_c = 100$ pF, $R_d = 800$ kΩ, $C_0 = 500$ fF, $L = 435$ nH and $Z_0 = 50$ Ω

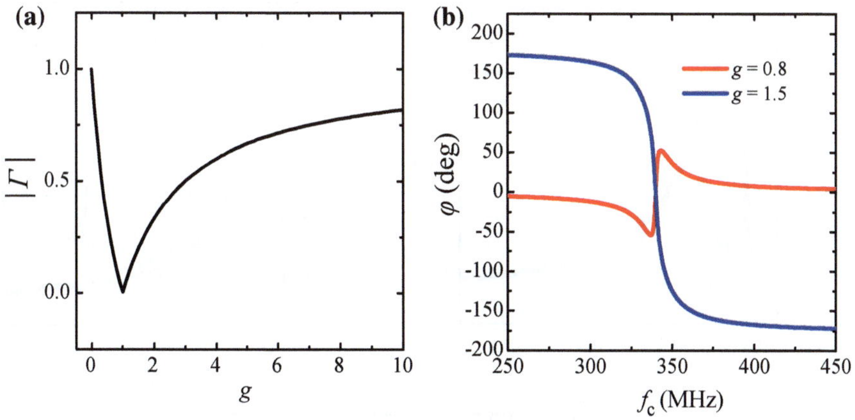

Fig. 12 **a** Simulation for the magnitude $|\Gamma|$ as a function of coupling g. **b** The phase $\varphi = \arg(\Gamma)$ as a function of rf carrier frequency f_c for different values of coupling constant g

$$\Delta\Gamma = -i\frac{4g}{\omega_L(1+g)^2}Q_0\Delta\omega_L. \tag{59}$$

The differential change in the loaded resonant frequency $\Delta\omega_L$ for changes in capacitance can be written as

$$\Delta\omega_L = -\frac{1}{2}\frac{1}{C_0}\omega_L\frac{\omega_L^2}{\omega_0^2}\Delta C_0. \tag{60}$$

Now using this expression for $\Delta\omega_L$ from Eq. (60), we can rewrite Eq. (59) as

$$\Delta\Gamma = i\frac{2g}{(1+g)^2}\cdot Q_0\cdot\frac{1}{(C_c+C_0)}\Delta C_0. \tag{61}$$

The coupling constant g can be evaluated at the loaded resonant frequency as

$$\begin{aligned} g\big|_{\omega\to\omega_L} &= Z_0C_c^2\omega_L^2(R_d+Z_0) \\ &= Z_0C_c^2\frac{(R_d+Z_0)}{L(C_c+C_0)} \\ &= \frac{Z_0}{R_{eq}}, \end{aligned} \tag{62}$$

where $R_{eq} = L(C_c+C_0)/R_dC_c^2$, is the equivalent resistance of the circuit at $f = f_{res}$, as depicted in Fig. 10b. We use this expression for g from Eq. (62) to rewrite Eq. (61) as

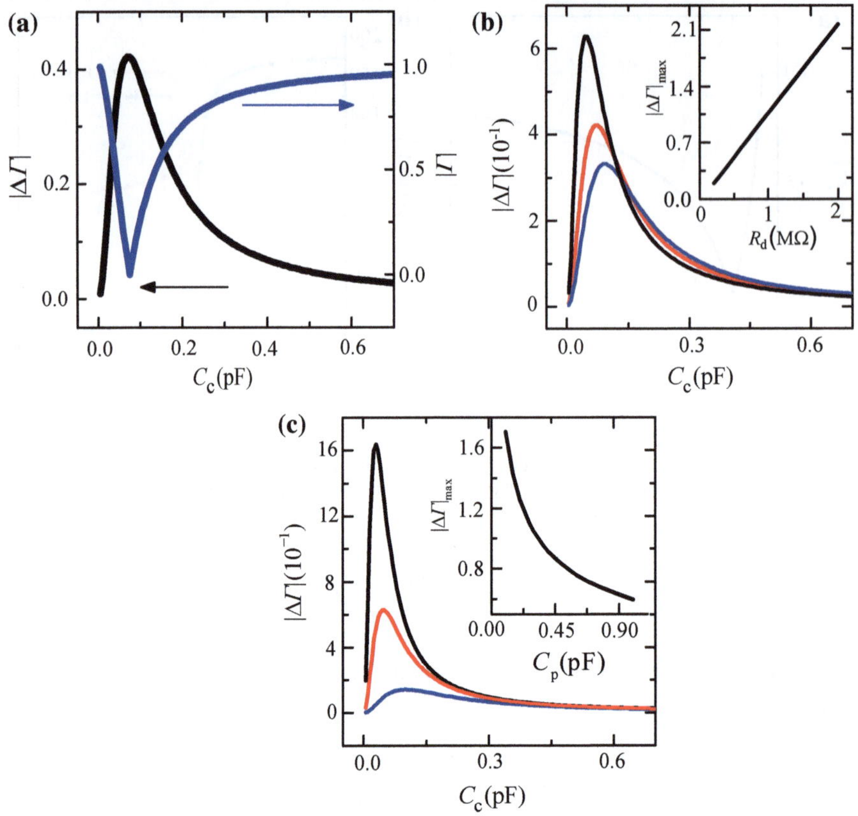

Fig. 13 Resonator Optimisation [26]. **a** $|\Delta\Gamma|$ (black) and $|\Gamma|$ (blue) as a function of C_c for the experimental values $R_d = 800\,\text{k}\Omega$, $C_p = 0.48$ pF and $L = 405$ nH, $Z_0 = 50\,\Omega$. We consider $\Delta C =$ 1 fF. **b** Dependence on R_d. $|\Delta\Gamma|$ as a function of C_c for $R_d = 200\,\text{k}\Omega$ (blue), 800 kΩ (red) and 2 MΩ (black). Inset: Maximum $|\Delta\Gamma|$ as a function of R_d. **c** Dependence on C_p. $|\Delta\Gamma|$ as a function of C_c for $C_p = 0.8$ pF (blue), 0.48 pF (red) and 0.2 pF (black). Inset: Maximum $|\Delta\Gamma|$ as a function of C_p

$$|\Delta\Gamma| = \frac{2R_{eq}Z_0}{(R_{eq}+Z_0)^2} Q_0 \frac{\Delta C_0}{(C_c + C_0)}. \tag{63}$$

Equation (63) provides the guidelines to optimize the sensitivity of gate-based sensing approach. The charge sensitivity δq is inversely proportional to $|\Delta\Gamma|$ and hence a study of this magnitude yields an estimate of the relative sensitivity level [69, 70].

Firstly, exemplified in Fig. 13a, where both $|\Delta\Gamma|$ and $|\Gamma|$ as a function of C_c are plotted, we observe that $|\Delta\Gamma|$ is maximum when the coupling capacitor is chosen to give perfect matching ($|\Gamma| = 0$) i.e. $R_{eq} = Z_0$. Secondly, increasing R_d, and in turn the unloaded Q-factor Q_0 (see Eq. 55), leads to an increase in sensitivity. The effect on $|\Delta\Gamma|$ of increasing R_d can be seen in Fig. 13b where the maximum $|\Delta\Gamma|$ increases as R_d is increased from 200 kΩ (blue) to 2 MΩ (black). Note the change in optimal C_c as R_d is varied. $|\Delta\Gamma|_{max}$ increases linearly with R_d as can be seen in the inset. Thirdly, a reduction of the circuit capacitance, by reducing C_p, leads to an enhanced sensitivity. This can be observed in Fig. 13c where we plot $|\Delta\Gamma|$ as a function of C_c and fixed L and R_d for three different values of C_p, 0.2 pF (black), 0.48 pF (red) and 0.8 pF (blue). $|\Delta\Gamma|_{max}$ decreases as $C_p^{-1/2}$, as can be seen in the inset. Finally, the change in capacitance, ΔC_0, needs to be maximized. This can be achieved by maximizing the gate coupling factor α which has a quadratic effect on $|\Delta\Gamma|$, see Eq. (22).

In short, the key requirements for sensitive capacitive gate-based readout are:

1. **Large Q_0 resonators,**
2. **Operating close to perfect matching condition g $\approx$ 1,**
3. **Low parasitic capacitance,**
4. **Large gate coupling to the quantum system.**

Note that large Q_0 combined with g $\approx$ 1 imply large Q_L is required (see Eq. 58) for sensitive gate-based sensing. Also Eq. (63) can be expressed in much simpler terms when the resonator is matched to the line $|\Delta\Gamma| = \pi R_d f_{res} \Delta C_0$. This is evident from the fact that $g = 1$ implies $Z_0 = R_{eq}$ in Eq. (62). In this case, one can see that the optimal device should have as low dissipation as possible (large R_d). Moreover, gate-based sensing benefits from operating at high frequency, assuming the the loss R_d stays fixed with increasing frequency.

5 How to Achieve Sensitive Gate-Based Sensing

In this section we discuss how the key requirements for capacitive gate-based sensing can be implemented experimentally.

5.1 Implementing Large Gate Coupling

The complementary metal-oxide-semiconductor (CMOS) technology has been optimizing the gate coupling to the channel of the transistors for decades to allow fast and reliable control over the logic circuits. The CMOS microelectronics industry went from 2D planar to 3D vertical nanowire (NW) geometry where the gate electrode wraps around the NW channel from multiple sides providing better electrostatic control over the channel [71]. We leverage this large gate-coupling by using a 300 mm CMOS platform compatible tri-gate silicon nanowire field effect transistors (NWFETs), see Fig. 14a. The silicon NW channel is doped with phosphorous at a concentration of 5×10^{17} cm^{-3}. The NW channel length $l = 30$ nm, width $w = 60$ nm and height $h = 11$ nm . The gate stack consists of 0.8 nm SiO_2 followed

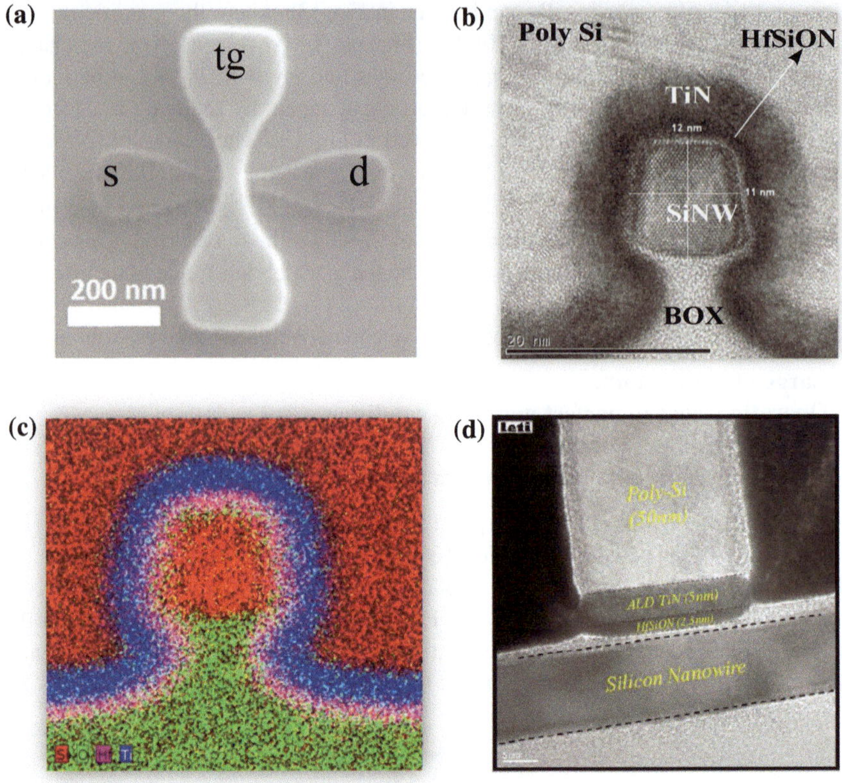

Fig. 14 **a** A scanning-electron micrograph of a NWFET showing source (s), drain (d) and top-gate (tg) terminals. **b** Transmission electron micrograph (TEM) of a Si NWFET cross section perpendicular to direction from s to d reservoir. **c** TEM image with energy dispersive X-ray spectroscopy (EDX) mapping. **d** TEM image of a Si NWFET cross section along the direction from s to d reservoir. Image courtesy of Dr. Sylvain Barraud, CEA-LETI, Grenoble, France

by 1.9 nm HfSiON. The HfSiON is capped by 5 nm TiN and 50 nm polycrystalline silicon, see Fig. 14b. These materials are identified in Fig. 14c using energy dispersive X-ray spectroscopy (EDX) mapping in a transmission electron microscope (TEM). The high-k dielectric gate stack provides better electrostatic control over the channel and smaller gate leakage. The transistor's multigate geometry, combined with a small equivalent gate oxide thickness of 1.3 nm, results in QDs with large gate couplings $\alpha = C_{tg}/C_{\Sigma} = 0.85 - 0.89$ since the total capacitance C_{Σ} is mostly given by its capacitance to the gate electrode C_{tg}. The TEM image of a cross section along the direction of transport is also shown in Fig. 14d. The 145 nm thick SiO_2 underneath the NW channel is known as the buried oxide (BOX). The 850 μm thick Si handle wafer underneath the BOX is phosphorus doped with doping level $N_d \approx 10^{15}\,cm^{-3}$. The handle wafer is used as a global back-gate (bg).

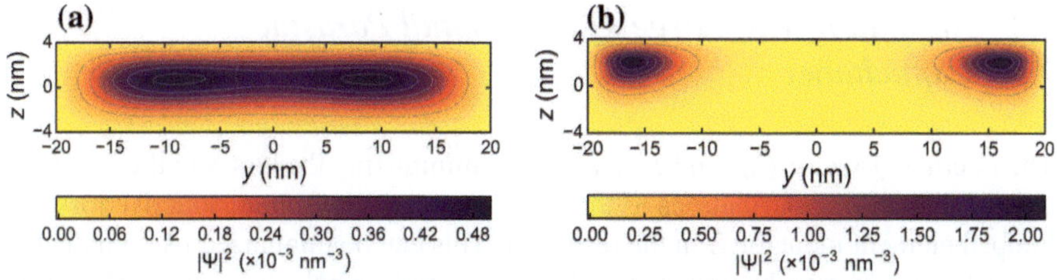

Fig. 15 **a** Simulation of the electron probability distribution $|\Psi|^2$ in the y-z cross section of the NW channel when $V_{bg} = -1$ V, showing the formation of a single QD and **b** for $V_{bg} = -10$ V, illustrating the double QD regime [73]

5.1.1 Tuning the QDs in NWFET

The global back-gate voltage can effectively tune the NWFET channel potential and controls the formation to QDs. The effect of the V_{bg} can be understood from simulating the electron probability distribution $|\Psi|^2$ in the NW channel using tight binding model [72, 73]. The existence of a single QD is evident as the $|\Psi|^2$ extends over the central part of the channel, see Fig. 15a, for $V_{bg} = -1$V. But at $V_{bg} = -10$ V, two dots are formed in parallel in the NW channel. Now, the electron probability is distributed over two regions located at the top-most corners, i.e. silicon corner dots [74, 75]. These simulations depend on device dimensions, remote Coulomb scattering and surface roughness.

5.1.2 Measuring the Gate Coupling

The gate coupling α can be measured from the cryogenic DC electrical transport measurements of the Si NWFET. The source-drain current I_{sd} is measured as function of top-gate voltage V_{tg}, back-gate voltage V_{bg} and source-drain voltage V_{sd}. The measurement set-up with the device schematic diagram is shown in Fig. 16. The charge stability diagram of the NWFET is characterized by measuring the source-drain current (I_{sd}) as function of V_{tg} and V_{bg} in the sub-threshold regime of device ($V_{tg} \sim V_{th}$), see Fig. 17a. At low temperatures, when the NWFET is biased below threshold ($V_{tg} \approx 0.5$V), few-electron QDs [1] are formed in the NW channel [74–76]. This measurement is performed for $V_{ds} = 0.1$ mV inside Kelvinox K400 dilution refrigerator. Transport in the NWFET is dominated by Coulomb blockade effect and I_{sd} exhibits Coulomb oscillations [2, 77]. Around $V_{bg} = 0$ V, the oscillations are quite regular and I_{sd} exhibits characteristic Coulomb diamonds when plotted as function of V_{tg} and V_{sd}, see Fig. 17b. The separation in V_{tg} between charge transition line I and II, ΔV_{tg}, at a given V_{sd} gives a measurement of $\alpha = V_{sd}/\Delta V_{tg}$. We repeat these measurements for several V_{sd} and obtain ΔV_{tg} as a function of V_{sd}, providing a measure of α from the slope and of the V_{sd} offset from the intercept. We obtain $\alpha = 0.85 \pm 0.01$.

5.2 Resonator with Large Q_0 and Small Parasitic Capacitance

The internal Q-factor Q_0 can be enhanced by minimizing the losses in the resonator. One of the major contributions to the total losses comes from conductive losses in lumped element inductor L in the resonator. This can be minimized using inductors made from superconducting materials, for example NbN [26]. The NbN can be deposited on sapphire substrate using DC magneton sputtering in a ultra high vacuum chamber ($\sim 10^{-9}$ mBar). The sputtered film can be characterized by X-ray diffraction (XRD) technique. The (111) XRD peak at $2\theta = 35.3°$, confirms NbN film as shown in Fig. 18a. This also indicates that the NbN film is (111)-textured as promoted by the c-plane sapphire substrate [78]. The (200) peak at $2\theta = 41.7°$ is shifted to a slightly higher angle compared to the NbN reference, which could be due to the presence of small amount of oxygen during the growth [79]. The superconducting transition or critical temperature T_c can be measured by monitoring the resistance of the NbN film using standard four point probe technique [80]. The resistance of the NbN film tends to zero below $T_c = 13.6$ K indicating the emergence of superconducting state in the NbN film, see Fig. 18b. The T_c is well above 4.2 K which makes the cryogenic characterization of the NbN possible in liquid He-4 bath. The measurements were performed in a liquid He-4 cryostat where the temperature can be controlled from 4.2 K to higher temperatures. The NbN spiral indicators can be fabricated using standard UV optical photolithography techniques and reactive ion etching (RIE) as shown in Fig. 19. In Fig. 20a, we have shown an optical image of a 26-turn spiral inductor. The reflection coefficient Γ can be measured by making a two port network which consists of the spiral inductor L and a capacitor C_{par} in parallel, see Fig. 20b. This LC parallel resonator is coupled to the external world by the coupling capacitor C_c. By measuring the magnitude of the reflection coefficient $|\Gamma|$ and fitting it with

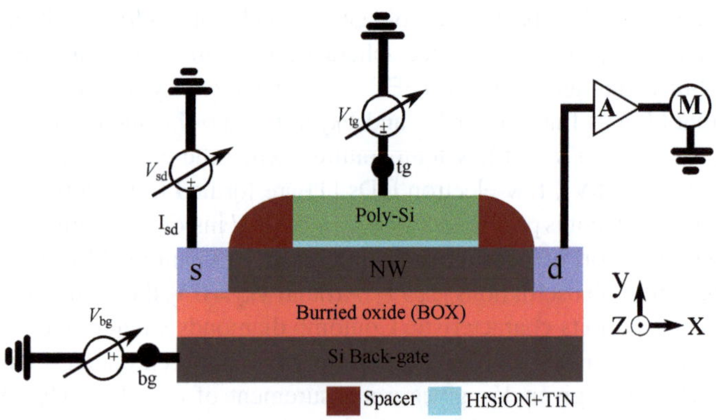

Fig. 16 Schematic diagram of DC measurement for single-gate Si NWFET. Externally applied bias voltages for top-gate (tg), back-gate (bg) and source-drain (sd) terminals are denoted by V_{tg}, V_{bg} and V_{sd}. Source-drain current I_{sd} is measured as a function of different bias voltages. Symbol A and M stand for current pre-amplifier and multi-meter respectively

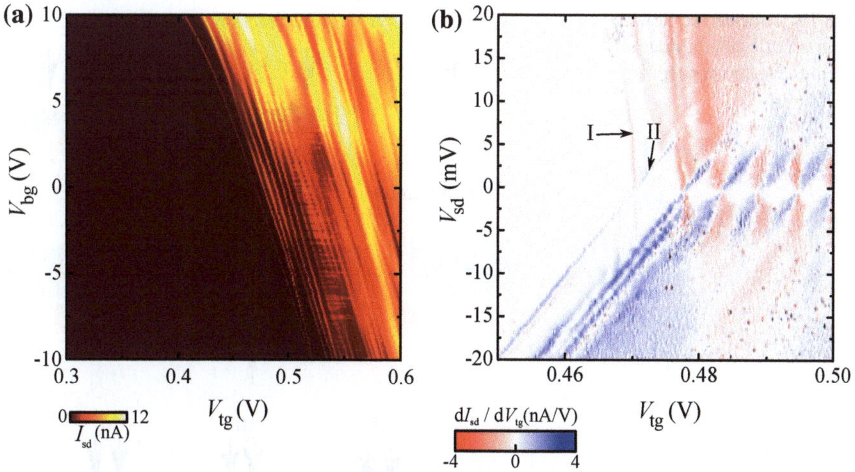

Fig. 17 **a** Source-drain current I_{sd} as a function of top-gate voltage V_{tg} and back-gate voltage V_{bg}. **b** Source-drain current I_{sd} as a function of top-gate voltage V_{tg} and source-drain voltage V_{sd} showing Coulomb diamonds

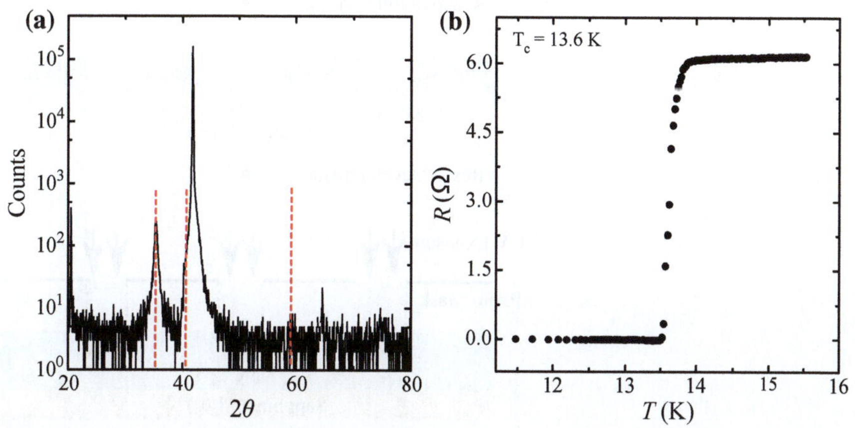

Fig. 18 **a** X-ray diffraction data for NbN film grown on sapphire (in black) and expected peak position for (111) and (200) planes. **b** NbN film resistance R as a function of temperature T

a Lorentzian line-shape, one can extract the resonant frequency f_{res} and the loaded Q-factor Q_L, see Fig. 20c. By repeating the Q_L for NbN inductors with different turn number n, it has been demonstrated that the loaded Q-factor $Q_L \approx 450$, see Fig. 20d [45].

The spiral NbN inductors also provide smaller parasitic capacitance as compared to wire bound lumped inductors, this primarily because of smaller footprints of the spiral on the sapphire substrate [21, 26, 81–83]. Also the sapphire between the spirals can be over etched to reduce the dielectric constant, minimizing the inter winding capacitance of the spiral.

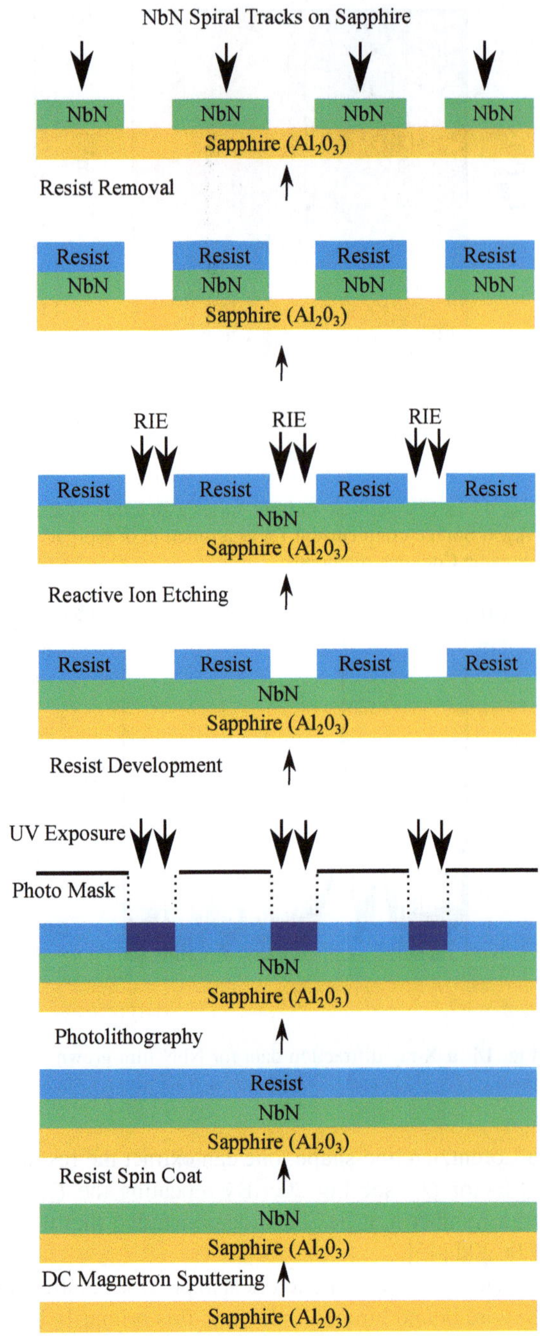

Fig. 19 Fabrication flow diagram of NbN circular spiral inductor on sapphire

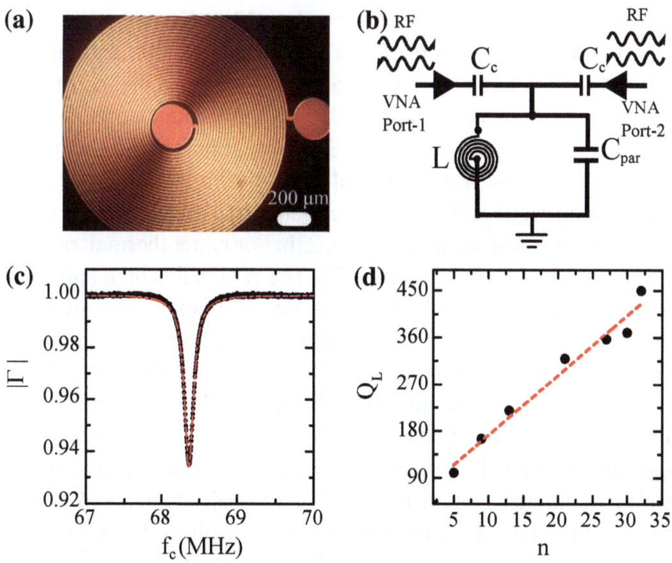

Fig. 20 **a** Optical image of $n = 26$-turn NbN spiral inductor. **b** Two-port network used to measure the Γ for a parallel resonator consisting of the spiral NbN inductor L, a capacitor $C_{\mathrm{par}} = 10$ pF. Γ is measured with vector network analyzer (VNA) using port 1. **c** Magnitude $|\Gamma|$ as a function of frequency f_c. **d** Loaded quality factor Q_{L} as a function of inductor turn number n (Data in black dot, linear fit in red dot)

5.3 *Operating Close to Perfect Matching*

The level of coupling between the resonator and the external line is characterized by the coupling constant $g = Q_0/Q_{\mathrm{e}}$. The g controls the power transfer between the resonator and external circuitry. A coupling constant $g = 1$ implies the matching condition is fulfilled and equal amount of power is dissipated in resonator and the external circuit. When $g < 1 (g > 1)$, one operates in an undercoupled (overcoupled) regime where more (less) power is dissipated in the resonator than in the external circuits. To operate close to perfect matching, one needs to tweak the coupling capacitor C_{c}, so that $g \sim 1$ [84].

6 Cryogenic RF Reflectometry Setup

Capacitive gate-based sensing uses rf reflectometry measurements at the base temperature of a dilution refrigerator. Here we introduce this technique from a hardware perspective of the Kelvinox K400 dilution refrigerator along with its different subsystems. In rf reflectometry, the rf carrier is supplied from a high performance low noise rf source (Rohde & Schwarz SMB 100A). The rf carrier splits into two signals passing through a directional coupler (Mini-Circuits ZEDC-15-2B), see Fig. 21. The signal from the coupled port, which is -15 dB attenuated, is directed through UT-114 Cu-Cu semi-rigid to the SMA input port at the top of the insert. Depending on the desired power level in the device, more room temperature attenuation (usually -20 to -30 dB) can be used at SMA input port at the top of the Kelvinox K400 insert. Inside the insert the rf signal travels from the input SMA port through UT-085 SS-SS semi-rigid and further attenuated by -20 dB with low temperature attenuator from Huber and Shuner, see Fig. 22b. For better thermalization, the SS-SS semi-rigid is bolted to 1 K plate, see Fig. 22. The rf signal is then split by ZEDC-15-2B directional coupler and the -15 dB attenuated coupled port signal reaches the mixing chamber (MC) through 250–550 MHz band pass filters. The semi-rigid line is thermalized with sapphire microstrip anchored on the MC plate. From MC plate the rf signal reaches the device through UT-085 Cu-Cu semi-rigid. The device acts as a modulator of the incident rf carrier and some part of the rf-carrier is reflected due to impedance mismatch at the gate of the device. The reflected rf signal travels through the unattenuated port of the MC directional coupler and reaches the still plate where it is thermalized again with sapphire stripline. The cryogenic amplifier Quinstar QCA-U350-30H is thermally anchored at the 4 K plate. The amplifier provides a gain $G \sim 40$ dB (bandwidth BW = 300–400 MHz and noise temperature $T_n = 4.5$ K at 350 MHz). The QCA-U350-30H is powered by FET servo switching power supply. The amplified signal reaches to SMA output port at the top of the insert and is further amplified at room temperature using two amplifiers Quinstar QCA-U350-30H (gain 40 dB) and Mini Circuits ZRL-700+ (gain 30 dB) in cascade. The ZRL-700+ is biased using Agilent E3631A power supply. After the amplification signal is demodulated using AD0540B I/Q demodulator. The demodulation is performed by multiplying the signal with the rf-carrier from the unattenuated port of the room temperature directional coupler. This produces base-band signals; the in-phase (I) and the quadrature (Q) phase signals, at the demodulator output. This technique is referred as the standard homodyne detection. Both I and Q signals are amplified by low noise preamplifier SR560 from Stanford Research before being recorded by LeCroy WaveRunner 44Xi-A or Teledyne LeCroy DHO4054A oscilloscopes, see Fig. 21.

7 Capacitive Gate-Based Sensing: Experiment

In this section, we present capacitive gate-based sensor measurements using the rf reflectometry technique [21, 29, 85, 86]. The quantum system considered here is a single QD in the NW channel tunnel coupled to the reservoir. Electron can tunnel back and forth between the coupled QD-reservoir as discussed in the Sect. 2.2. The resonator is formed by wirebonding the NbN inductor in parallel to the device gate capacitance to ground, C_d, and the circuit parasitic capacitance, C_p, and coupled to the Z_0 line via a coupling capacitor C_c as shown in Fig. 23a. We estimate $L = 405$ nH,

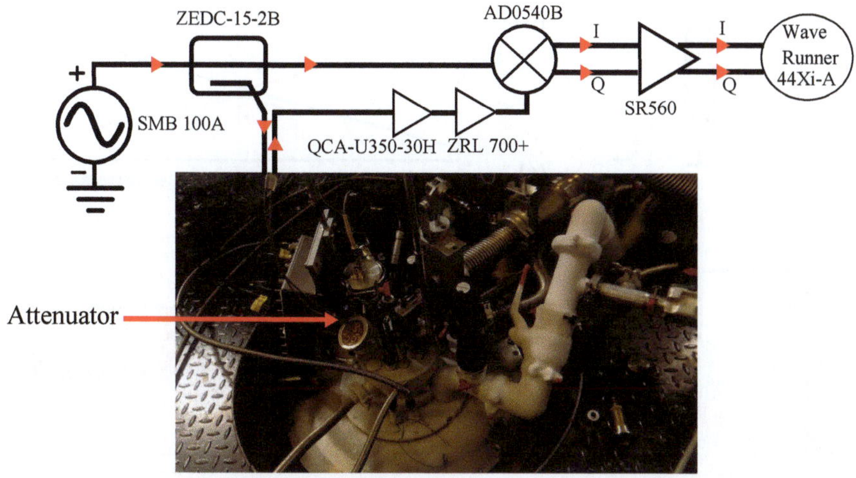

Fig. 21 Room temperature electronic set-up for homodyne detection in rf reflectometry measurements

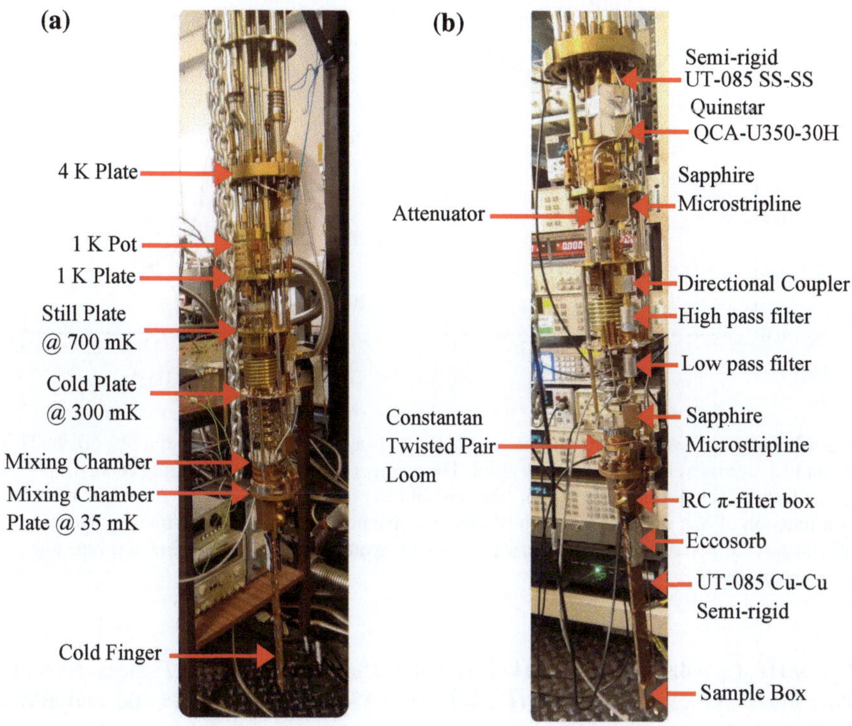

Fig. 22 **a** Insert of the Kelvinox K400 dilution fridge showing different temperature stages. **b** DC and rf components and wiring of the insert

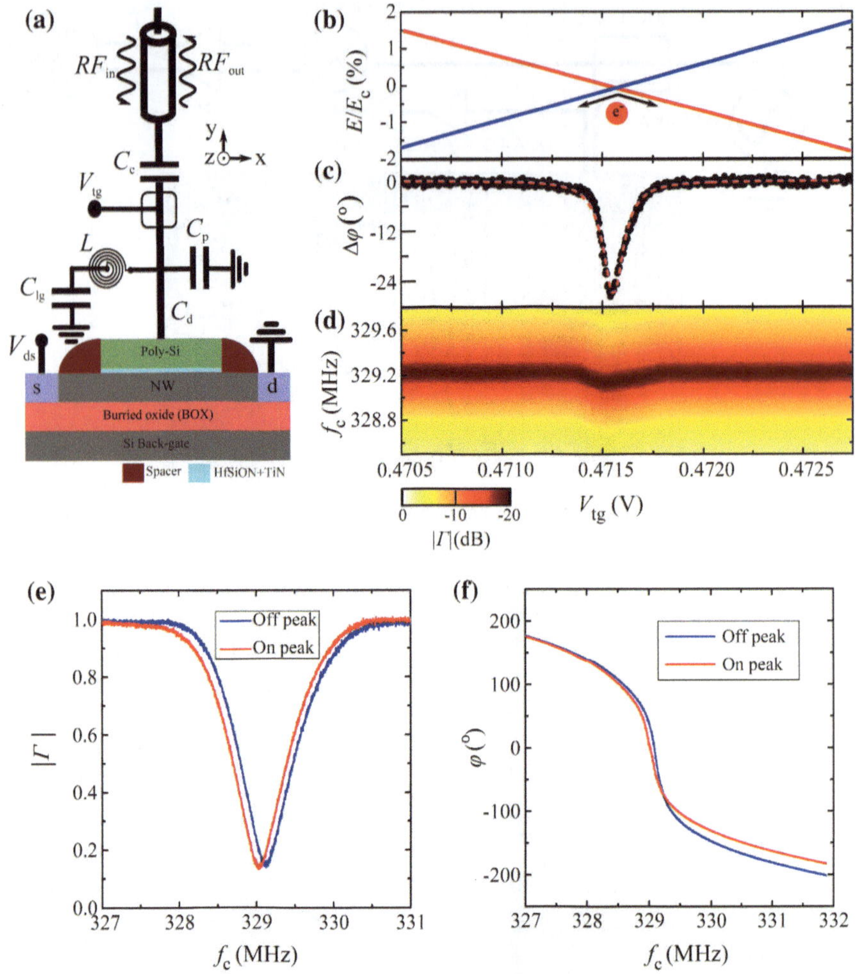

Fig. 23 Dispersive regime [26]. **a** The gate based sensor. **b** Energy diagram of a fast driven TLS, E_0 and E_1, across a charge degeneracy point. The energy is normalised to charging energy E_c of the QD. **c** Data (black dot) and Lorentz fit (red dashed line) for the phase change $\Delta\varphi$ of the resonator as a function of V_{tg}. **d** $|\Gamma|$ as function of carrier frequency f_c and V_{tg}. Experimental data for $|\Gamma|$ (**e**) and ϕ (**f**) at two different V_{tg} voltages—away from and at the charge degeneracy, blue and red traces respectively

$C_c = 90$ fF, $C_p = 480$ fF, $R_d = 800$ kΩ from the reflection coefficient Γ measurements. This gives us $f_{res} = 1/(2\pi\sqrt{L(C_c + C_0)}) = 329.33$ MHz, $Q_L \approx 400$ and BW= 0.82 MHz [26].

7.1 Dispersive Regime

The resonator is coupled to a single QD in the NW channel, its susceptance changes when adiabatic single-electron tunneling occurs between the QD and the s/d reservoirs. At resonance, variations in φ capture changes in the device capacitance ΔC that can be attributed to tunneling capacitance C_t [31]. Since the resonator is overcoupled, $\Delta\varphi = -2Q_L\Delta C/(C_c + C_0)$. The origin of ΔC can be explained by considering an uncoupled two level system (TLS) described by a QD with zero (E_0) or one excess electron (E_1) as shown in Fig. 23b, see Sect. 2.2. The resulting tunneling capacitance is given by

$$C_t = \frac{(\alpha e)^2}{\pi} \frac{h\gamma}{(h\gamma)^2 + (\alpha e \Delta V_{tg})^2}, \tag{64}$$

where $\Delta V_{tg} = V_{tg} - V_{tg}^0$ and the V_{tg}^0 is the top-gate voltage at the charge degeneracy point. [23, 44]. This voltage-dependent C_t produces a phase shift in the resonator as can be seen in Fig. 23c where $\Delta\varphi$ is plotted as the V_{tg} is swept across the charge degeneracy point. The maximum phase shift $\Delta\varphi = 28°$ at the degeneracy point. Using Eq. (64), $\Delta\varphi$ is fitted (red curve) and $\gamma = 26$ GHz ($\gg f_{res}$) is extracted from the full-width-half-maximum (FWHM). C_t loads the resonator pulling down its resonant frequency as shown in Fig. 23d. Finally, in Fig. 23d, e, $|\Gamma|$ and φ are compared on and off the degeneracy point. From these measurements dispersive shift $\Delta f = 88$ kHz has been extracted, which corresponds to an effective change in capacitance given by $\Delta C = 2(C_c + C_0)\Delta f / f_{res} = 0.3$ fF. This agrees well with the expected maximum tunneling capacitance 0.37 fF, calculated from Eq. (64).

7.2 Charge Sensitivity Measurement

We use the technique in Ref. [27] to measure the charge sensitivity of the capacitive gate-based sensor: A small sinusoidal voltage is applied to the top-gate of the device with a root mean square charge equivalent amplitude Δq and frequency f_m. This produces an amplitude modulation of the carrier that results in sidebands appearing in the power spectrum of the reflected signal at $f_c \pm f_m$, as can be seen in Fig. 24 along with the rf carrier. The height of the sideband measured from the noise floor defines the power signal-to-noise ratio (SNR). The charge sensitivity is then calculated from the definition, $\delta q = \Delta q/(\sqrt{2\text{RBW}} \times 10^{\frac{\text{SNR}}{20}})$ where RBW is the resolution bandwidth of the spectrum analyzer [87]. A separate figure of merit is the modulation depth dBc, given by the sideband height relative to the carrier in dB. This figure indicates how much of the input signal is modulated by the device. Figure 24 shows the power spectrum at the optimal working point of the resonator (identified by measurements described below). The spectrum is obtained using a modulation signal with $f_m = 511$ Hz, $\Delta q = 6.98 \times 10^{-4} e$ and a RBW $= 10$ Hz. From the mea-

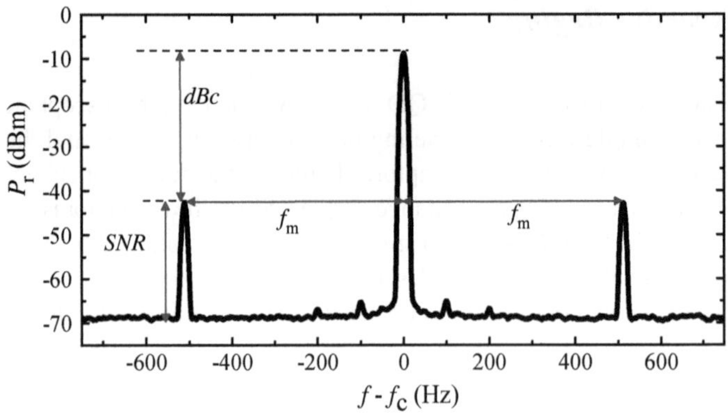

Fig. 24 Charge sensitivity measurement. Reflected power P_r spectrum showing the sidebands at $f_c \pm f_m$ due to gate-voltage modulation [26]

sured SNR=26.1 dB, we estimate $\delta q = 7.7\ \mu e/\sqrt{Hz}$. This result represents a charge sensitivity improvement of a factor of ~5 with respect to previous reports [22].

7.3 Gate Sensor Optimization

Now we discuss how the optimal gate-sensor working point in terms of V_{tg}, f_c and carrier power P_c can be obtained. In Fig. 25a, the lower sideband SNR as a function of V_{tg} and P_c is shown. For a fixed P_c, the SNR shows two maxima whose separation in V_{tg} increases as P_c is increased. The position of the maxima corresponds to the V_{tg} points of maximum slope on either side of the charge transition in Fig. 23b. The dependence of the separation with increasing P_c indicates that the transition is being broadened by the rf voltage V_c. Only at the lowest values of P_c, where the separation between peaks remains constant, the transition is lifetime limited. To find the optimal carrier power for sensing, the maximum SNR at each P_c is obtained (black dots) as shown in Fig. 25b. The SNR peaks at $P_c = -98$ dBm when the transition is still power broadened, as opposed to the expectation that the maximum SNR would be achieved when the transition is lifetime broadened. To understand why, lets define the modulation index $M = 10^{\frac{dBc}{20}}$ and note that the sideband SNR can be expressed as SNR$\propto M \times P_c$. M decreases as the rf carrier voltage V_c increases as seen (black dots) in Fig. 25c. This dependence can be modelled as a convolution of two competing processes: lifetime broadening and power broadening. The transition is lifetime broadened by γ to produce a linewidth V_γ ($h\gamma = e\alpha V_\gamma$) and power broadened by applied rf carrier power with a linewidth proportional to V_c. The dependence of M on V_c can be approximated by $1/\sqrt{V_\gamma^2 + V_c^2}$ which decreases with increasing V_c as shown in Fig. 25c (red dashed curve). The maximum SNR occurs when the increase in input power is compensated by the decrease in M.

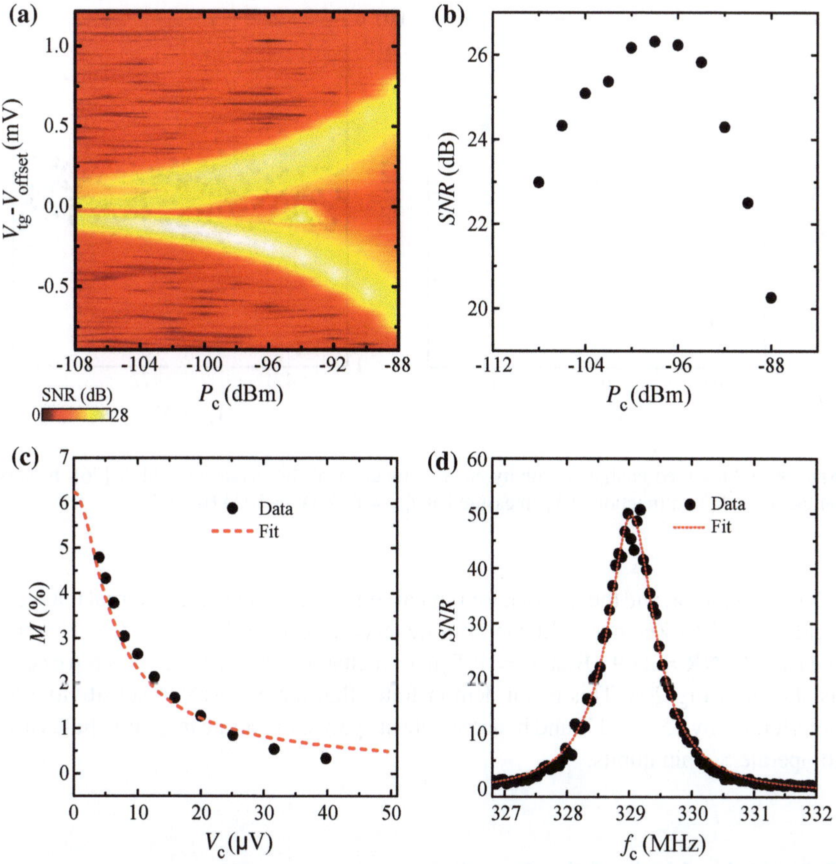

Fig. 25 Charge sensitivity Optimization [26] **a** SNR in dB as a function of V_{tg} and P_c (V_{offset} = 472.23 mV). **b** Sideband SNR as function of P_c. **c** Modulation index M (black dots) and fit (red dashed line) as a function of input rf-carrier voltage V_c. **d** Sideband power SNR in linear scale versus rf-carrier frequency f_c measured at $P_c = -115$ dBm

To find the optimal carrier frequency, lets plot the sideband power SNR (in linear scale) at $f_c - f_m$ as a function of rf carrier frequency f_c when swept across a frequency range containing f_{res}, Fig. 25d. The SNR shows a Lorentzian profile with center frequency f_{res}, and BW and Q_L. To avoid power broadening, the input RF power was kept at -115 dBm during this measurement.

7.4 *Gate Sensor in Magnetic Field*

The sensor's charge sensitivity on in-plane magnetic field B is discussed here. This is important because of the use of a superconducting material (NbN) for the lumped

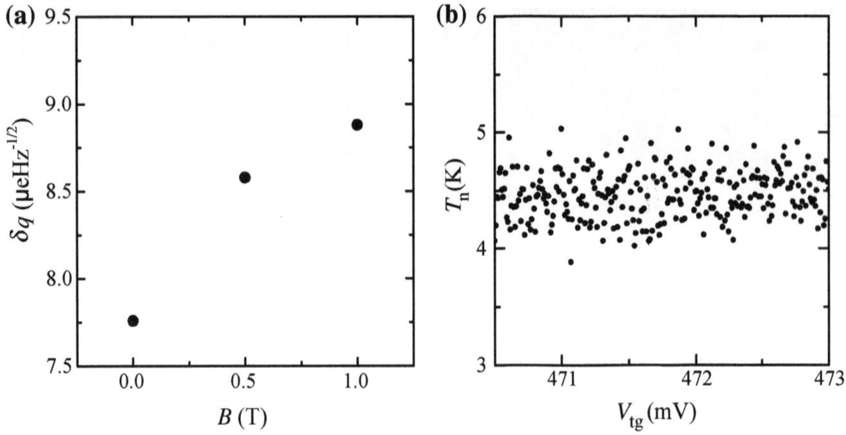

Fig. 26 **a** Measured charge sensitivity δq as a function of the magnetic field B [26]. **b** Noise temperature T_n as a function of V_{tg} measured at f_{res} with RBW = 300 kHz [26]

element inductor, and the fact that for typical spin qubit systems, an external magnetic field is used to Zeeman-split the spin degenerate energy levels [1, 15, 88]. The measured SNR = 24.9 dB at $B = 1$ T gives a charge sensitivity $\delta q = 8.8\ \mu e/\sqrt{\text{Hz}}$ as shown in Fig. 26a. This result demonstrates that the gate-sensor sensitivity only deteriorates by 15% at 1 T and hence is robust against moderate magnetic fields used to operate Si spin qubits.

7.5 *Noise Limit of the Gate Sensor*

Finally, in Fig. 26b, we show the noise temperature T_n of the system at the resonant frequency f_{res} as a function of V_{tg}. As the sweep V_{tg} is swept across the charge degeneracy point, T_n stays constant at 4.5 ± 0.4 K which matches with the noise temperature of the cryogenic amplifier (Quinstar QCA-U350-30H) in the measurement setup. This implies that charge sensitivity is limited by the thermal noise of the cryo-amplifier and not by the Sisyphus noise which can be orders of magnitude smaller [22].

7.6 *High Frequency Operation*

To asses the advantage of operating at higher frequencies, we present measurements for capacitive gate-based sensor at $f_{res} = 616.18$ MHz. We use a NbN inductor with $L = 134$ nH. In Fig. 27a, b, we present the magnitude $|\Gamma|$ and phase ϕ of the reflec-

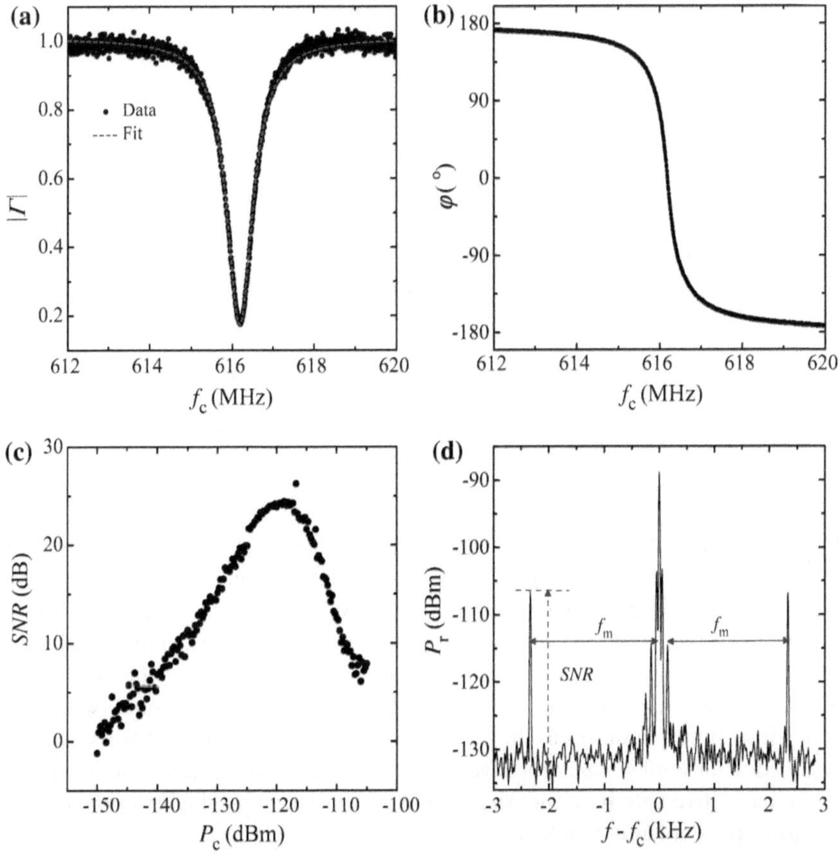

Fig. 27 Resonator characterization from reflection coefficient measurement [26]. **a** Magnitude $|\Gamma|$ of the reflection coefficient Γ measured (in black) and fit (in red dashed line). **b** Phase $\varphi = \arg(\Gamma)$. **c** Sideband SNR as a function of P_c. **d** Reflected power P_r spectrum showing the sidebands at $f_c \pm f_m$ due to gate-voltage modulation

tion coefficient. The resonator has a $BW = 0.78$ MHz and hence a loaded Q-factor $Q_L = 790$, is close to matching $|\Gamma|_{min} = 0.1$ and overcoupled $Q_e < Q_0$. The optimized charge sensitivity of 1.3 $\mu e/\sqrt{Hz}$ at $P_c = -120$ dBm is obtained. The SNR is peaked at -120 dBm as depicted in Fig. 27c. The reflected power P_r spectrum shows the sidebands at $f_c \pm f_m$ due to gate-voltage modulation in Fig. 27d. For this measurement $\Delta q = 1.48 \times 10^{-4} e$ and RBW = 20 Hz. The sensitivity improvement is 30 times when compared to previous reports [22] and places the charge sensitivity of the capacitive gate-based sensors on a par with the best reported rf-SET sensitivities [87, 89]. In this case, the energy sensitivity $\delta\varepsilon$ is $36\hbar$ since the charging energy $E_c = 28$ meV [89].

Moreover the advantage of the improved sensitivity is demonstrated by acquiring a charge stability map of the device containing 512×256 data points in just 700 ms,

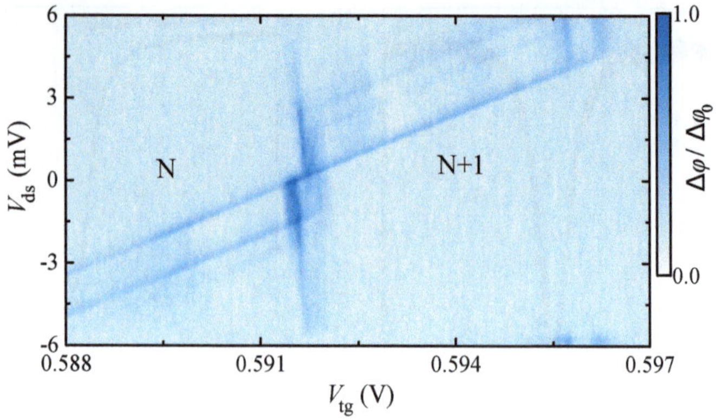

Fig. 28 Fast data acquisition of a dot to reservoir transition. V_{ds} is ramped at 4 kHz while V_{tg} at 7 Hz. Each trace is averaged 5 times. The total measurement time is 700 ms [26]

see Fig. 28. A double ramp scheme [90] is used where drain voltage, V_{ds} ramps up at 4 kHz (sawtooth) while slowly ramp V_{tg} at 7 Hz (triangular). The frequency of the ramp is limited in this measurement by the low-pass filtering in our lines (cut-off frequency 5 kHz). The 2D map is composed of traces of the demodulated phase response. In Fig. 28 the characteristic signature of Coulomb blockade is measured dispersively with the gate-based sensor. The combination of the data quality and short acquisition time demonstrates the potential of this new gate-based sensor design for fast readout of semiconductor nanostructures. Given the integration time of 5 μs per point, capacitive gate-based sensing may enable performing single-shot readout of electron spin dynamics in silicon.

8 Conclusion

We have presented capacitive gate-based sensing in Si QD using rf reflectometry techniques. The charge sensitivity of the capacitive sensor is comparable to the best sensitivities ever demonstrated for the rf-SET [87]. In the case of the rf-SET the experimental sensitivity is limited by shot-noise whereas in the case of our gate-sensor is limited by the noise of the cryogenic amplifier. Hence by using a quantum-limited amplifier like Josephson parametric amplifier at the mixing chamber of a dilution refrigerator ($\approx$10 mK) instead of a more standard cryogenic amplifier ($T_n \approx 4K$), one can get a reduction of the charge sensitivity making the gate-sensor the most sensitive electrometer [91]. However the ultimate sensitivity of this dispersive sensor remains to be explored and a study should consider the effects of the Sysiphus noise [22] and the Johnson-Nyquist noise of the resonator which can be orders of magnitude lower than shot-noise at miliKelvin temperatures and radio-frequencies. Additionally, the charge sensitivity demonstrated combined with the bandwidth of the resonators results in a few microsecond detection with an rms charge noise ($\Delta Q = \delta q \times \sqrt{\mathrm{BW}}$) well below 1 e. Therefore, capacitive gate-based sensing allows single-shot readout of electron spins in silicon double QDs [92–95] where the relaxation and coherence times are typically larger than a microsecond [96, 97].

References

1. R. Hanson, L.P. Kouwenhoven, J.R. Petta, S. Tarucha, L.M.K. Vandersypen, Rev. Mod. Phys. **79**, 1217 (2007)
2. F.A. Zwanenburg, A.S. Dzurak, A. Morello, M.Y. Simmons, L.C.L. Hollenberg, G. Klimeck, S. Rogge, S.N. Coppersmith, M.A. Eriksson, Rev. Mod. Phys. **85**, 961 (2013)
3. L. Fricke, M. Wulf, B. Kaestner, F. Hohls, P. Mirovsky, B. Mackrodt, R. Dolata, T. Weimann, K. Pierz, U. Siegner et al., Phys. Rev. Lett. **112**, 226803 (2014)
4. T. Ihn, S. Gustavsson, U. Gasser, B. Küng, T. Müller, R. Schleser, M. Sigrist, I. Shorubalko, R. Leturcq, K. Ensslin, Solid State Commun. **149**, 1419 (2009)
5. S. Gustavsson, R. Leturcq, T. Ihn, K. Ensslin, M. Reinwald, W. Wegscheider, Phys. Rev. B. **75**, 075314 (2007)
6. D. McClure, L. DiCarlo, Y. Zhang, H.-A. Engel, C. Marcus, M. Hanson, A. Gossard, Phys. Rev. Lett. **98**, 056801 (2007)
7. A. Rossi, T. Ferrus, D.A. Williams, Appl. Phys. Lett. **100**, 133503 (2012)
8. D. Maradan, L. Casparis, T.-M. Liu, D.E.F. Biesinger, C.P. Scheller, D.M. Zumbühl, J.D. Zimmerman, A.C. Gossard, J. Low. Temp. Phys. **175**, 784 (2014)
9. A. Mavalankar, S.J. Chorley, J. Griffiths, G.A.C. Jones, I. Farrer, D.A. Ritchie, C.G. Smith, Appl. Phys. Lett. **103**, 133116 (2013)
10. I. Ahmed, A. Chatterjee, S. Barraud, J.J.L. Morton, J.A. Haigh, M.F. Gonzalez-Zalba, Commun. Phys. **1**, 66 (2018a)
11. T.F. Watson, S.G.J. Philips, E. Kawakami, D.R. Ward, P. Scarlino, M. Veldhorst, D.E. Savage, M.G. Lagally, M. Friesen, S.N. Coppersmith, M.A. Eriksson, L.M.K. Vandersypen, Nature **555**, 633 (2018)
12. F.H.L. Koppens, J.A. Folk, J.M. Elzerman, R. Hanson, L.H.W. Van Beveren, I.T. Vink, H.-P. Tranitz, W. Wegscheider, L.P. Kouwenhoven, L.M.K. Vandersypen, Science **309**, 1346 (2005)
13. J.R. Petta, A.C. Johnson, J.M. Taylor, E.A. Laird, A. Yacoby, M.D. Lukin, C.M. Marcus, M.P. Hanson, A.C. Gossard, Science **309**, 2180 (2005)
14. M. Veldhorst, H.G.J. Eenink, C.H. Yang, A.S. Dzurak, Nat. Commun. **8**, 1766 (2017)
15. A. Morello, J.J. Pla, F.A. Zwanenburg, K.W. Chan, K.Y. Tan, H. Huebl, M. Mottonen, C.D. Nugroho, C. Yang, J.A. van Donkelaar, A.D.C. Alves, D.N. Jamieson, C.C. Escott, L.C.L. Hollenberg, R.G. Clark, A.S. Dzurak, Nature **467**, 687 (2010)
16. K.D. Petersson, L.W. McFaul, M.D. Schroer, M. Jung, J.M. Taylor, A.A. Houck, J.R. Petta, Nature **490**, 380 (2012)
17. A. Stockklauser, P. Scarlino, J.V. Koski, S. Gasparinetti, C.K. Andersen, C. Reichl, W. Wegscheider, T. Ihn, K. Ensslin, A. Wallraff, Phys. Rev. X **7**, 011030 (2017)
18. X. Mi, J.V. Cady, D.M. Zajac, P.W. Deelman, J.R. Petta, Science **355**, 156 (2017)
19. N. Samkharadze, G. Zheng, N. Kalhor, D. Brousse, A. Sammak, U.C. Mendes, A. Blais, G. Scappucci, L.M.K. Vandersypen, Science **359**, 1123 (2018)
20. X. Mi, M. Benito, S. Putz, D.M. Zajac, J.M. Taylor, G. Burkard, J.R. Petta, Nature **555**, 599 (2018)
21. J.I. Colless, A.C. Mahoney, J.M. Hornibrook, A.C. Doherty, H. Lu, A.C. Gossard, D.J. Reilly, Phys. Rev. Lett. **110**, 046805 (2013)
22. M.F. Gonzalez-Zalba, S. Barraud, A.J. Ferguson, A.C. Betz, Nat. Commun. **6**, 6084 (2015)
23. M.G. House, T. Kobayashi, B. Weber, S.J. Hile, T.F. Watson, J. Van Der Heijden, S. Rogge, M.Y. Simmons, Nat. Commun. **6**, 8848 (2015)

24. N. Ares, F.J. Schupp, A. Mavalankar, G. Rogers, J. Griffiths, G.A.C. Jones, I. Farrer, D.A. Ritchie, C.G. Smith, A. Cottet, G.A.D. Briggs, E.A. Laird, Phys. Rev. Appl. **5**, 034011 (2016)
25. A. Crippa, R. Maurand, D. Kotekar-Patil, A. Corna, H. Bohuslavskyi, A.O. Orlov, P. Fay, R. Laviéville, S. Barraud, M. Vinet, M. Sanquer, S.D. Franceschi, X. Jehl, Nano Lett. **17**, 1001 (2017)
26. I. Ahmed, J.A. Haigh, S. Schaal, S. Barraud, Y. Zhu, C.-M. Lee, M. Amado, J.W.A. Robinson, A. Rossi, J.J.L. Morton, M.F. Gonzalez-Zalba, Phys. Rev. Appl. **10**, 014018 (2018b)
27. R.J. Schoelkopf, P. Wahlgren, A.A. Kozhevnikov, P. Delsing, D.E. Prober, Science **280**, 1238 (1998)
28. R.C. Ashoori, H.L. Stormer, J.S. Weiner, L.N. Pfeiffer, S.J. Pearton, K.W. Baldwin, K.W. West, Phys. Rev. Lett. **68**, 3088 (1992)
29. C. Ciccarelli, A.J. Ferguson, New J. Phys. **13**, 093015 (2011)
30. F. Persson, C.M. Wilson, M. Sandberg, G. Johansson, P. Delsing, Nano Lett. **10**, 953 (2010)
31. R. Mizuta, R.M. Otxoa, A.C. Betz, M.F. Gonzalez-Zalba, Phys. Rev. B **95**, 045414 (2017)
32. D.V. Averin, K.K. Likharev, J. Low Temp. Phys. **62**, 345 (1986)
33. T.A. Fulton, G.J. Dolan, Phys. Rev. Lett. **59**, 109 (1987)
34. K. Barnham, D. Vvedensky, *Low-Dimensional Semiconductor Structures: Fundamentals and Device Applications* (Cambridge University Press, 2008)
35. S. Shin, J. Lee, H. Kang, J.B. Choi, S.-R.E. Yang, Y. Takahashi, D. Hasko, Nano lett. **11**, 1591 (2011)
36. T. Ihn, *Semiconductor Nanostructures* (Oxford University Press, Oxford, 2010)
37. L.P. Kouwenhoven, D.G. Austing, S. Tarucha, Rep. Prog. Phys. **64**, 701 (2001)
38. L.P. Kouwenhoven, C.M. Marcus, P.L. McEuen, S. Tarucha, R.M. Westervelt, N.S. Wingreen, Proc. NATO Adv. Study Inst., 105 (1997)
39. C.W.J. Beenakker, Phys. Rev. B, **44**, 1646 (1991)
40. C. Wasshuber, H. Kosina, Superlattices Microstruct. **21**, 37 (1997)
41. C.W.J. Beenakker, H. van Houten, in *Solid State Physics*, Vol. 44 (Elsevier, 1991), pp. 1–228
42. M.A. Kastner, Phys. Today **46**, 24 (1993)
43. R.C. Ashoori, Nature **379**, 413 (1996)
44. A. Cottet, C. Mora, T. Kontos, Phys. Rev. B **83**, 121311 (2011)
45. I. Ahmed, *Radio-Frequency Capacitive Gate-based Sensing for Silicon CMOS Quantum Electronics*, Ph.D. thesis, University of Cambridge (2019)
46. D. Loss, D.P. DiVincenzo, Phys. Rev. A **57**, 120 (1998)
47. J. Gorman, D.G. Hasko, D.A. Williams, Phys. Rev. Lett. **95**, 090502 (2005)
48. W.G. Van der Wiel, S. De Franceschi, J.M. Elzerman, T. Fujisawa, S. Tarucha, L.P. Kouwenhoven, Rev. Mod. Phys. **75**, 1 (2002)
49. A.C. Betz, R. Wacquez, M. Vinet, X. Jehl, A.L. Saraiva, M. Sanquer, A.J. Ferguson, M.F. Gonzalez-Zalba, Nano Lett. **15**, 4622 (2015)
50. F. Hofmann, T. Heinzel, D.A. Wharam, J.P. Kotthaus, G. Böhm, W. Klein, G. Tränkle, G. Weimann, Phys. Rev. B **51**, 13872 (1995)
51. M.F. Gonzalez-Zalba, *Single Donor Detection in Silicon Nanostructures*, Ph.D. thesis, University of Cambridge (2013)
52. M. Larsson, *Electron Transport in Quantum Dots Defined in Low-Dimensional Semiconductor Structures*, Ph.D. thesis (2011)
53. A.K. Hüttel, S. Ludwig, H. Lorenz, K. Eberl, J.P. Kotthaus, Phys. Rev. B **72**, 081310 (2005)
54. M. Pioro-Ladriere, M.R. Abolfath, P. Zawadzki, J. Lapointe, S.A. Studenikin, A.S. Sachrajda, P. Hawrylak, Phys. Rev. B **72**, 125307 (2005)
55. M.F. González-Zalba, A. Saraiva, M. Calderón, D. Heiss, A.J. Koiller, B. Ferguson, Nano lett. **14**, 5672 (2014)
56. S.J. Hile, M.G. House, E. Peretz, J. Verduijn, D. Widmann, T. Kobayashi, S. Rogge, M.Y. Simmons, Appl. Phys. Lett. **107**, 093504 (2015)
57. https://iopscience.iop.org/article/10.1088/1367-2630/14/2/023050
58. A. Wallraff, D.I. Schuster, A. Blais, L. Frunzio, J. Huang, R.S. Majer, S. Kumar, S.M. Girvin, R.J. Schoelkopf, Nature **431**, 162 (2004)

59. T. Frey, P. Leek, M. Beck, J. Faist, A. Wallraff, K. Ensslin, T. Ihn, M. Büttiker, Phys. Rev. B **86**, 115303 (2012)
60. D. Kim, D.R. Ward, C.B. Simmons, J.K. Gamble, R. Blume-Kohout, E. Nielsen, D.E. Savage, M.G. Lagally, M. Friesen, M.A. Coppersmith, Eriksson, Nat. Nanotechnol. **10**, 243 (2015)
61. M.F. Gonzalez-Zalba, S.N. Shevchenko, S. Barraud, J.R. Johansson, A.J. Ferguson, F. Nori, A.C. Betz, Nano Lett. **16**, 1614 (2016)
62. M.A. Sillanpää, L. Roschier, P.J. Hakonen, App. Phys. Lett. **87**, 092502 (2005)
63. T. Duty, G. Johansson, K. Bladh, D. Gunnarsson, C. Wilson, P. Delsing, Phys. Rev. Lett. **95**, 206807 (2005)
64. D. Berns, W. Oliver, S. Valenzuela, A. Shytov, K. Berggren, L. Levitov, T. Orlando, Phys. Rev. Lett. **97**, 150502 (2006)
65. C. Tahan, R. Joynt, Phys. Rev. B **89**, 075302 (2014)
66. O. Astafiev, Y.A. Pashkin, Y. Nakamura, T. Yamamoto, J.-S. Tsai, Phys. Rev. Lett. **93**, 267007 (2004)
67. K. Wang, C. Payette, Y. Dovzhenko, P. Deelman, J.R. Petta, Phys. Rev. Lett. **111**, 046801 (2013)
68. P. Boross, G. Széchenyi, A. Pályi, Nanotechnology **27**, 314002 (2016)
69. L. Roschier, P. Hakonen, K. Bladh, P. Delsing, K.W. Lehnert, L. Spietz, R.J. Schoelkopf, J. Appl. Phys. **95**, 1274 (2004)
70. T. Müller, T. Choi, S. Hellmüller, K. Ensslin, T. Ihn, S. Schön, Rev. Sci. Inst. **84**, 083902 (2013)
71. I. Ferain, C.A. Colinge, J.P. Colinge, Nature **479**, 310 (2011)
72. Y.M. Niquet, D. Rideau, C. Tavernier, H. Jaouen, X. Blase, Phys. Rev. B **79**, 245201 (2009)
73. D.J. Ibberson, L. Bourdet, J.C. Abadillo-Uriel, I. Ahmed, S. Barraud, M.J. Calderón, Y.M. Niquet, M.F. Gonzalez-Zalba, Appl. Phy. Lett. **113**, 053104 (2018)
74. B. Voisin, V.H. Nguyen, J. Renard, X. Jehl, S. Barraud, F. Triozon, M. Vinet, I. Duchemin, Y.M. Niquet, S. de Franceschi, M. Sanquer, Nano Lett. **14**, 2094 (2014)
75. H. Sellier, G.P. Lansbergen, J. Caro, S. Rogge, N. Collaert, I. Ferain, M. Jurczak, S. Biesemans, Appl. Phys. Lett. **90**, 073502 (2007)
76. A.C. Betz, S. Barraud, Q. Wilmart, B. Placais, X. Jehl, M. Sanquer, M.F. Gonzalez-Zalba, Appl. Phys. Lett. **104**, 043106 (2014)
77. Y. Nagamune, H. Sakaki, L.P. Kouwenhoven, L.C. Mur, C.J. P.M. Harmans, J. Motohisa, and H. Noge, Appl. Phys. Lett. **64**, 2379 (1994)
78. J.C. Powder Diffraction Standards, on Joint committee on powder diffraction standards, 9 (1967)
79. S. Ran, L. Gao, J. Am. Ceram. Soc. **90**, 2626 (2007)
80. D.K. Schroder, *Semiconductor Material and Device Characterization* (Wiley, New York, 2006)
81. T.R. Stevenson, F.A. Pellerano, C.M. Stahle, K. Aidala, R.J. Schoelkopf, Appl. Phys. Lett. **80**, 3012 (2002)
82. J.M. Hornibrook, J.I. Colless, A.C. Mahoney, X.G. Croot, S. Blanvillain, H. Lu, A.C. Gossard, D.J. Reilly, Appl. Phys. Lett. **104**, 103108 (2014)
83. G. Stojanovic, L.J. Zivanov, in *International Conference on Microelectronics*, Vol. 2 (IEEE, 2004), pp. 613–616
84. https://aip.scitation.org/doi/full/10.1063/1.5082894
85. S.J. Chorley, J. Wabnig, Z.V. Penfold-Fitch, K.D. Petersson, J. Frake, C.G. Smith, M.R. Buitelaar, Phys. Rev. Lett. **108**, 036802 (2012)
86. K.D. Petersson, C.G. Smith, D. Anderson, P. Atkinson, G.A.C. Jones, D.A. Ritchie, Nano Lett. **10**, 2789 (2010)
87. H. Brenning, S. Kafanov, T. Duty, S. Kubatkin, P. Delsing, J. Appl. Phys. **100**, 114321 (2006)
88. M. Veldhorst, C.H. Yang, J.C.C. Hwang, W. Huang, J.P. Dehollain, J.T. Muhonen, S. Simmons, A. Laucht, F.E. Hudson, K.M. Itoh, A. Morello, A.S. Dzurak, Nature **526**, 410 (2015)
89. A. Aassime, G. Johansson, G. Wendin, R.J. Schoelkopf, P. Delsing, Phys. Rev. Lett. **86**, 3376 (2001)
90. J. Stehlik, Y.-Y. Liu, C.M. Quintana, C. Eichler, T.R. Hartke, J.R. Petta, Phys. Rev. Applied **4**, 014018 (2015)

91. https://journals.aps.org/prl/accepted/45074Y00Td815f8fb07c0695224d14d130011b382
92. A. West, B. Hensen, A. Jouan, T. Tanttu, H. Yang, C.A. Rossi, M.F. Gonzalez-Zalba, F. Hudson, A. Morello, D.J. Reilly, A.S. Dzurak, Nat. Nanotechnol. **14**, 437–441 (2019)
93. M. Urdampilleta, D.J. Niegemann, E. Chanrion, B. Jadot, C. Spence, P.-A. Mortemousque, L. Hutin, B. Bertrand, S. Barraud, R. Maurand, X. Jehl, V.M. Ranceschi, S. De, T. Meunier, Nature Nanotechnol
94. G. Zheng, N. Samkharadze, M.L. Noordam, N. Kalhor, D. Brousse, A. Sammak, G. Scappucci, L.M.K. Vandersypen, arXiv preprint arXiv:1901.00687 (2019)
95. P. Pakkiam, A.V. Timofeev, M.G. House, M.R. Hogg, T. Kobayashi, M. Koch, S. Rogge, M.Y. Simmons, Phys. Rev. X **8**, 041032 (2018)
96. M. Veldhorst, J.C.C. Hwang, C.H. Yang, A.W. Leenstra, B. de Ronde, J.P. Dehollain, J.T. Muhonen, F.E. Hudson, K.M. Itoh, A. Morello, A.S. Dzurak, Nat. Nanotechnol. **9**, 981 (2014)
97. E. Kawakami, P. Scarlino, D.R. Ward, F.R. Braakman, D.E. Savage, M.G. Lagally, M. Friesen, S.N. Coppersmith, M.A. Eriksson, L.M.K. Vandersypen, Nat. Nano. **9**, 666 (2014)
98. M.D. Stewart, N.M. Zimmerman, Appl. Sci. **6**, 187 (2016)

Micro Electrometers Based on Micromachined Time-Modulated Variable Capacitors

Joshua E.-Y. Lee

Abstract This chapter describes the concept of a micro electrometer based on a time-modulated variable capacitor. The roots of the device concept trace back to a period prior to the availability of microfabrication technology and techniques. This chapter shows how microfabrication technology, when aptly applied to older concepts that were previously realized using conventional machining techniques, can provide significant advancements in performance. This is particularly the case where the parasitic elements at the interface between parts of the system have an impact on the sensor performance. Through tighter integration of constituent parts, significant improvements in performance can be realized through further miniaturization.

Micro and nano electrometers are instruments for measuring precise quantities of charge, where the key device has a size in the micro- or nano- scale. Such instruments have been realized through two approaches. One approach is in the form of solid-state devices, a classic example of which is the single-electron transistor (SETs). While SETs provide unrivalled charge resolution, they typically are cryogenically cooled for operation. The other approach is in the form of mechanical devices based on oscillating structures. The charge that is applied to the input of such a mechanical device can be sensed either through a frequency modulated output or an amplitude modulated output. In the frequency modulated configuration, the amount of charge is measured through the perturbation of the oscillating structure's natural frequency. In the amplitude modulated configuration, the amount of charge is measured by an amplitude change in the output. This chapter focuses on the amplitude modulation approach wherein the oscillation amplitude remains unperturbed by the input charge. The micro electrometer approach described in this chapter is based on the vibrating reed electrometer proposed in the 1930s by Gunn [1]. The approach has merits in terms of the operation temperature compared to some nanoscale implementations of the frequency modulated approach that have to operate at a temperature of a few Kelvins. The device concept had been proposed prior to the widescale availability of microfabrication technology with the advent of the microelectronics

J. E.-Y. Lee (✉)
Department of Electrical Engineering, City University of Hong Kong, Kowloon, Hong Kong
e-mail: josh.lee@cityu.edu.hk

Y. Zhu (ed.), *Micro and Nano Machined Electrometers*,
https://doi.org/10.1007/978-981-13-3247-0_4

industry. With more mature development of micromachining techniques to fabricate mechanical structures at the microscale, Riehl developed the concept of the vibrating reed electrometer by Gunn as a micromechanical device at the turn of the millennium. Riehl had implemented the concept in a polysilicon surface micromachining process, which was one of the earliest micromachining processes to have been developed. The polysilicon micromachining process was instrumental to the first commercial accelerometers. As the vibrating reed concept was based a capacitive structure, the polysilicon micromachining process had a drawback of offering limited structural thickness and thus limited capacitive overlapping area. Nonetheless, Riehl was able to demonstrate a charge resolution of 51 electrons/$\sqrt{\text{Hz}}$ [2]. One only has to compare this performance with the specifications of a commercially available electrometer to appreciate the level of improvement in performance with such micro electrometers. A state-of-art commercial electrometer would have a resolution in the order of 10fC. The subsequent development of bonded silicon technology [or silicon-on-insulator (SOI)], made available structural thicknesses that could be much thicker than what was feasibly possible with the polysilicon micromachining process. Subsequent work based on bonded silicon micromachining technology was able to demonstrate a charge resolution of 6 electrons/$\sqrt{\text{Hz}}$, thus approaching the limit of single electron resolution. In the design and conceptualization of the type of micro electrometer described in this chapter, considerations of both the mechanical aspects of the device and the electronics of the interface circuit have to be weighed together. In the rest of the chapter, various aspects on the variable capacitor electrometer are presented, starting from the basic concept, the effect of the fabrication process on performance, and the interface electronics.

1 Device Concept

The charge (Q) stored on a capacitor (C) can be determined by measuring the associated voltage across the capacitor, such that:

$$V = Q/C \tag{1}$$

Detecting a static charge stored on such a capacitor requires measuring a direct current (DC) voltage. For measuring precise amounts of charge, the voltage on the capacitor must be amplified using electronic amplifiers. Figure 1 shows the typical noise spectrum present in modern electronics where the low end of the frequency spectrum is dominated by flicker (i.e. 1/f) noise while the higher end of the frequency spectrum is dominated by white noise. The crossover point where 1/f noise meets white noise is known as the corner frequency, as depicted in Fig. 1 This figure illustrates the challenge of detecting DC signals, as the noise level at DC is highest relative to higher frequencies. The comparatively high level of flicker noise at DC therefore degrades the resolution of the charge sensor (i.e. electrometer). As such, it is advantageous to up modulate the DC voltage output on the capacitor to shift its

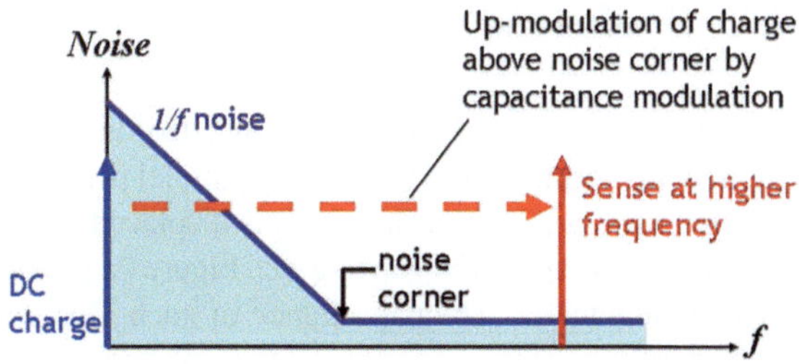

Fig. 1 Frequency spectrum of noise typical of detection electronics illustrating the motivation for up modulating the DC input to a higher frequency above the noise corner into the white noise region where the noise level is much lower

frequency above the corner frequency where white noise dominates and the noise levels are much lower, thus enhancing the sensor resolution.

Converting a static charge to an alternating current (AC) voltage would require modulating the capacitance of the charge sensor in the form a time-modulated variable capacitor. Using a modulating frequency of *f*, the output voltage from the variable capacitor will have a frequency component at *f*. Such a time-modulated variable capacitor could be implemented by varying the capacitor gap with time, whereby the plates of the capacitor are displaced relative to the other. The actuation mechanism to drive these movable plates is described in the next section.

2 Actuator Design

One of the most straightforward ways to realize an actuator in the silicon-based micromachining process is to etch capacitive structures. Actuation through these capacitive structures is realized through the electrostatic force when applying a DC voltage (V_{dc}) and AC voltage (modulated at an angular frequency $\omega = 2\pi f$) across the capacitor. The resulting electrostatic force (F) is given by:

$$\begin{aligned} F &= \frac{1}{2}\frac{dC}{dx}(V_{dc} + V_{ac}\cos(\omega t))^2 \\ &= \frac{1}{2}\frac{dC}{dx}\left[V_{dc}^2 + \frac{1}{2}V_{ac}^2 + 2V_{dc}V_{ac}\cos(\omega t) + V_{ac}^2\cos(2\omega t)\right] \end{aligned} \quad (2)$$

As we can see from Eq. (2), the generated electrostatic force has three frequency components: one at DC, one at the modulation frequency (*f*), and another at twice the modulation frequency (i.e. 2*f*). In the case of the variable capacitor electrometer, only the force component at *f* is desired, while the components at DC and 2*f* should be suppressed as far as possible.

In the following, we consider various common capacitive actuator topologies before concluding on the option most suitable for the case of the variable capacitor electrometer.

2.1 Parallel Plate

We first consider a parallel plate capacitive actuator, where the force and motion applied are normal to the plates. The parallel plate capacitor has an unperturbed gap denoted by g, and one of the plates is displaced by a distance x, thus increasing the instantaneous size of the gap to $g + x$. Figure 2a provides a schematic a parallel plate capacitive actuator. The capacitance of such a parallel plate capacitive actuator is given by:

$$C = \frac{\varepsilon_0 A}{g + x} \tag{3}$$

where ε_0 is the permittivity in vacuum (which is approximately close to the value in air) and A is the overlapping area between the pair of plates (i.e. $A = WL$ in the case of Fig. 2a). As the force that is generated from a capacitive actuator depends on the differential change in capacitance (dC/dx), according to Eq. (3), we thus consider the dC/dx term for the parallel plate capacitor:

$$\frac{dC}{dx} = -\frac{\varepsilon_0 A}{(g + x)^2} = -\frac{\varepsilon_0 A}{g^2}\left(1 - \frac{x}{g} + O\left(x^2\right)\right) \tag{4}$$

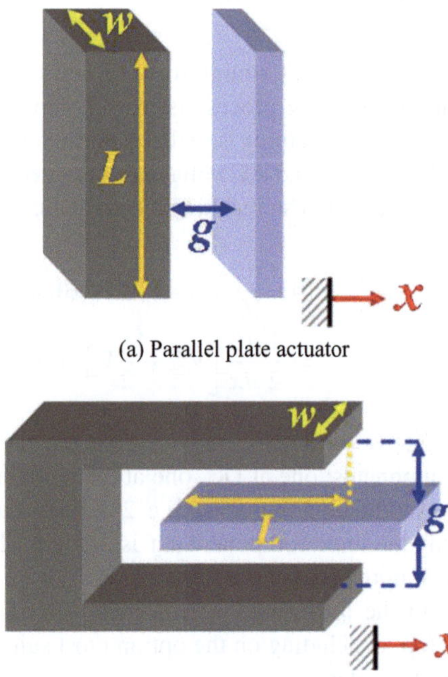

(a) Parallel plate actuator

(b) Comb drive actuator

Fig. 2 Schematic of typical capacitive actuators; **a** parallel plate capacitive actuator, **b** comb drive capacitive actuator. Note that the plates on the left side are fixed while the plates on the right are movable

The above equation has been obtained using a Taylor series expansion up to the second order term. As can be seen from Eq. (4), dC/dx (and thus also the force) for a parallel plate capacitive actuator is a function of the plate displacement save for the limit where $x \ll g$. In the case of the parallel plate capacitive actuator, the maximum plate displacement is limited by the capacitor gap. From Eq. (4), we see that narrowing the gap increases dC/dx, and thus increasing the efficiency of converting voltage into force. However, narrowing the gap simultaneously reduces the maximum allowable distance for the plate the move. When the plates are driven hard enough such that the plate displacement becomes comparable to gap, the electrostatic force becomes nonlinear (i.e. no longer independent of x). Consequently, nonlinear vibrations will arise in the oscillations of the variable capacitor, which should be avoided.

2.2 Comb Drive

Having seen that the parallel plate capacitive actuator is inherently nonlinear, we next consider the comb drive actuator. Figure 2b provides a schematic of a comb drive capacitive actuator. In the case of the comb drive capacitive actuator, the force and motion are applied parallel to the plates, such that the plates slide in and out relative to each other. The capacitance of a comb drive with N number of fingers or plates on one side is given by:

$$C = \frac{2N\varepsilon_0 W}{g}(L - x) \tag{5}$$

where L denotes the unperturbed overlapping length between the fingers, x is the displacement of the movable fingers, and W is the width of the fingers. Once again, we find the dC/dx term to evaluate the effect of the actuator topology on the associated electrostatic force generated:

$$\frac{dC}{dx} = -\frac{2N\varepsilon_0 H}{g} \tag{6}$$

In contrast to the parallel plate capacitive actuator, we see that the dC/dx term (and thus also the force) is independent of the displacement. Besides, the distance of travel of the movable plate is not limited by the gap unlike the parallel gap capacitor. With the aim of providing linear oscillations over reasonably large amplitudes of plate movement, the comb drive topology is suited for the case of the variable capacitor electrometer compared to the parallel plate capacitor.

However, choosing comb drive topology only ensures that dC/dx is independent of x. As seen from Eq. (2), further modifications are necessary to eliminate the DC and $2f$ component terms in the actuating force.

2.3 *Differential Comb Drive*

Applying a superposition of DC and AC voltages to a capacitive actuator on one side of the structure to be driven results in an additional DC component and 2*f* component. By applying the same DC voltage but reversing the phase of the AC voltage, the resulting electrostatic force is given by:

$$\begin{aligned} F &= \frac{1}{2}\frac{dC}{dx}(V_{dc} - V_{ac}\cos(\omega t))^2 \\ &= \frac{1}{2}\frac{dC}{dx}\left[V_{dc}^2 + \frac{1}{2}V_{ac}^2 - 2V_{dc}V_{ac}\cos(\omega t) + V_{ac}^2\cos(2\omega t)\right] \end{aligned} \quad (7)$$

As can be seen from Eq. (7), reversing the phase of the AC voltage keeps the DC and 2*f* components unchanged but reverses the phase of the component at *f*. Hence by placing a pair of comb drives back to back, such that the forces act opposite to each other, applying a differential pair of AC voltages results in cancellation of the DC and 2*f* components. This leaves only the force component at *f*, such that:

$$F = \frac{dC}{dx}[2V_{dc}V_{ac}\cos(\omega t)] \quad (8)$$

Therefore, a differential comb drive actuator design allows us to modulate the position of the time-modulated variable capacitor plates required for sensing charge in an electrometer. In principle, the only frequency component in the forcing term is the modulation frequency of the AC drive voltage (*f*).

3 Charge Feedthrough from Actuator Drive

Ideally, when there is no charge on the variable capacitor, the output voltage sensed should be zero. In practice, due to the presence of parasitic elements between the drive port and sense port of the device, some of the AC drive voltage applied at the drive port will be coupled into the sense port. In the case of electrometers, it is the parasitic capacitance that appears between the drive and sense ports, as depicted by the circuit model in Fig. 3, that is of most relevance in our analysis here. In Fig. 3, C_f is the parasitic capacitor while C_S is the capacitance of the variable capacitor. For a drive voltage of V_d (same as V_{ac}) the feedthrough voltage that develops across C_S is given by:

$$V_s = \frac{C_f}{C_s + C_f}V_d \quad (9)$$

As such, even without any charge applied on the variable capacitor electrometer, we can see how an output can be imposed by the drive voltages. As such, due to the presence of such parasitics, an offset is imposed by the drive input. It is worth noting

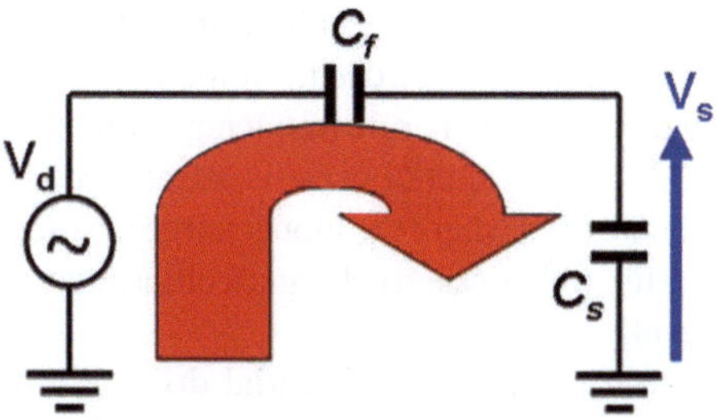

Fig. 3 Circuit model illustrating charge feedthrough from a single drive input to the to the variable capacitor (C_s) via a parasitic capacitor that appears between the drive and sense ports

that even though C_f is usually small (<1 pF), the typical drive voltage levels will be much higher than the minute voltage levels to be sensed across the variable capacitor associated with small input charges. As the variable capacitor electrometer senses an input charge by modulating it into an AC voltage, the offset imposed by the drive voltages appears as an equivalent charge imposed on the electrometer. As such, we here refer to the signal fed from the drive to the sense port as charge feedthrough.

By applying a pair of differential drive signals, as illustrated in Fig. 4, the level of feedthrough can be reduced substantially. The resulting feedthrough voltage that is seen across the variable capacitor is given by,

$$V_s = \left(\frac{C_{f1}}{C_{f1} + C_s} - \frac{C_{f2}}{C_{f2} + C_s}\right) V_d \tag{10}$$

By employing balanced structures where the placement of the two drive inputs are symmetrical with reference to the variable capacitor for sensing charge, C_{f1} and C_{f2} can be reasonable well-matched. If we denote the fractional mismatch between C_{f1} and C_{f2} by δ, then the feedthrough voltage sensed across the variable capacitor expressed as a function of δ,

$$V_s \approx \frac{C_f}{C_s} \delta V_d \tag{11}$$

where C_f is the average value between C_{f1} and C_{f2} [i.e. $C_f = 1/2\,(C_{f1} + C_{f2})$] and δ is defined as $\delta = |C_{f1} - C_{f2}|/C_f$. As seen from the above equation, the smaller the mismatch between the parasitic capacitors, the lower will be the associated charge feedthrough. In the hypothetical case where both parasitic feedthrough paths are perfectly identical equal (i.e. $C_{f1} = C_{f2}$), then the charge feedthrough would theoretically

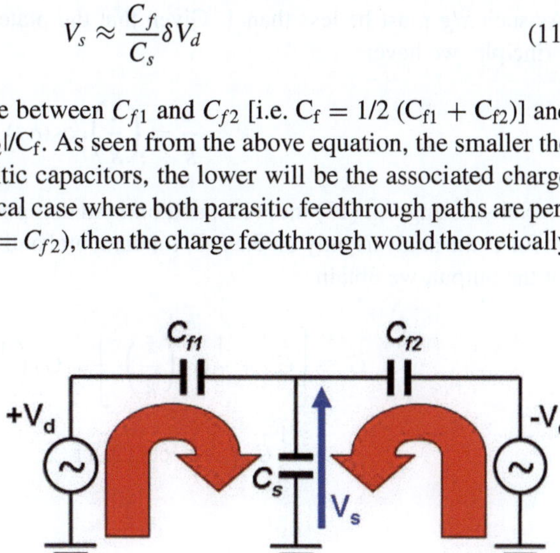

Fig. 4 Circuit model illustrating charge feedthrough from the drive inputs to the variable capacitor via parasitic elements in a differential drive setup

be zero. It is worth recalling from Sect. 2 that differential drive is advantageous in producing an electrostatic force that linear with the amplitude of AC drive voltage and contains only one frequency component (which is the drive frequency). Here we see that differential drive is advantageous for reducing charge feedthrough. In the classic case where the input charge was sensed at the modulation drive frequency (i.e. f), the offset due to charge feedthrough would be proportional to the drive voltage level.

But even with differential drive, the offset from charge feedthrough could still be very large in comparison to small amounts of input charges expected of single-electron resolution electrometers. In the following section, we explain how charge feedthrough can be further suppressed by several orders of magnitude using second harmonic detection.

4 Second Harmonic Detection

As has been described above, the variable capacitor electrometer may be formed by moving two parallel plates relative to each other. For example, the capacitance could be modulated by changing the capacitor gap. As such, the voltage-to-charge relation of a variable capacitor electrometer of any given capacitor topology can be expressed using a Maclaurin's series as a function of the plate displacement (x),

$$G = \frac{V}{Q} = \frac{1}{C(\kappa)} = G_o + G_1\kappa + G_2\kappa^2 + O\left(\kappa^3\right) \tag{12}$$

where κ is a dimensionless displacement variable defined by normalizing the plate displacement to the capacitor's nominal gap (g) such that $\kappa = x/g$. In this case, we assume that the plate displacement results in a change in the instantaneous gap. As such x/g must be less than 1. Given that the plate displacement is harmonic in principle, we have:

$$\kappa = \frac{x}{g} = \left(\frac{\bar{x}}{g}\right)\cos(\omega t) \tag{13}$$

Substituting Eq. (13) into (12), and omitted components above the 2nd harmonic to focus the analysis only on the DC, fundamental and second harmonic components of the output, we obtain,

$$G \approx \left[G_o + \frac{1}{2}G_2\left(\frac{\bar{x}}{g}\right)^2\right] + G_1\left(\frac{\bar{x}}{g}\right)\cos(\omega t) + \frac{1}{2}G_2\left(\frac{\bar{x}}{g}\right)^2\cos(2\omega t) \tag{14}$$

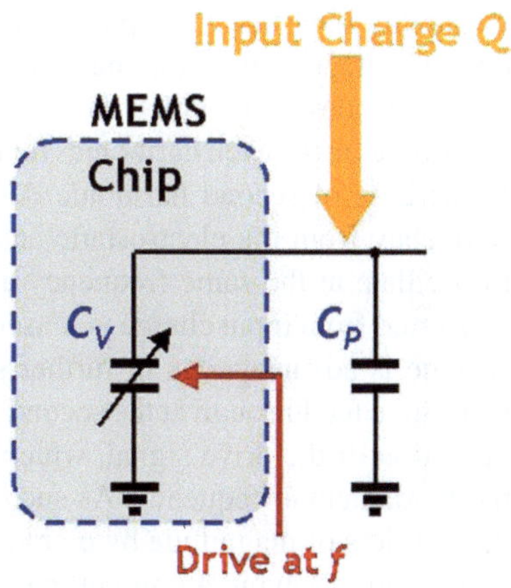

Fig. 5 Circuit model of time-modulated variable capacitor electrometer (C_V) with a shunt parasitic capacitor (C_P)

In deriving the voltage-to-charge relation for any given variable capacitor topology, we must include the shunt parasitic capacitance (C_P), as depicted in the circuit model shown in Fig. 5. While C_v is time-modulated, C_P is constant, such that,

$$G = \frac{1}{C_v(t) + C_P} \tag{15}$$

As such, the efficiency of the variable capacitor in up-modulating the input charge to an AC voltage is reduced by the presence of C_P. In the case of a simple parallel plate unit cell design where one plate moves relative to another that is fixed, the coefficients of the Maclaurin's series are,

$$\begin{aligned} G_0 &= \frac{1}{C_S + C_P} \\ G_1 &= \frac{C_S}{(C_S + C_P)^2} \\ G_2 &= \frac{-C_S C_P}{(C_S + C_P)^3} \end{aligned} \tag{16}$$

Here, C_S denotes the nominal capacitance of the unit cell when the movable plate is in its equilibrium position. Therefore, we can see that in the case of a simple parallel plate capacitor unit cell design, an input static charge will produce voltage outputs across the variable capacitor with components at the DC, *f* and *2f* (as well as all other higher harmonics). As such, by detecting the charge at the modulation frequency, a fraction of the comparably large AC drive voltage will be coupled to the variable capacitor to impose a large offset in the electrometer. We also see that for all frequency components of the output voltage, the larger the value of C_P relative to C_S, the lower the conversion from charge to voltage.

Having seen that a simple parallel plate capacitor unit cell design results in an output voltage with components at both even and odd harmonics, we consider an alternative design in which the odd harmonics are eliminated. This results in larger amplitudes at the even harmonics relative to the simple parallel plate capacitor design. In the case of second harmonic detection, the output is sensed at twice the drive frequency from the electrostatic actuations. In other words, the structure is driven to oscillate at the same frequency as the AC drive voltage (i.e. f), but the voltage converted from input charge is sensed at twice that frequency (i.e. $2f$). This detection scheme is advantageous in further suppressing charge feedthrough. The extent of capacitive feedthrough at the second harmonic is determined by the second harmonic distortion of the drive signal, which is typically in the order of −60 dB relative to the fundamental frequency. As such, charge feedthrough can be reduced further by three orders of magnitude by employing second harmonic detection.

Instead of having a unit cell based on one pair of parallel plates, we exploit the nonlinearity inherent to a balanced pair of back-to-back capacitors, as depicted in Fig. 6. In the unit cell of balanced capacitor structure shown in Fig. 6, the two plates on the left and right are fixed while the plate in between is moves between the two fixed plates. When the movable plate is in its equilibrium position, the gap relative to the left and right plates are equal, as also the capacitance. With the movable central plate grounded, the two side plates form a pair of identical capacitors in parallel (when the central plate is in the middle) with a total capacitance denoted by C_S. As the central plate moves towards the right, the gap with the plate on the right side is reduced by the extent of displacement (x) while the gap with the plate on the left side is increased by the same amount.

We can write down the instantaneous capacitance of the unit cell, including the parasitic shunt capacitance (C_P) as function of the central plate's displacement,

$$C(t) = \frac{C_S/2}{1-(x/g)} + \frac{C_S/2}{1+(x/g)} + C_P \tag{17}$$

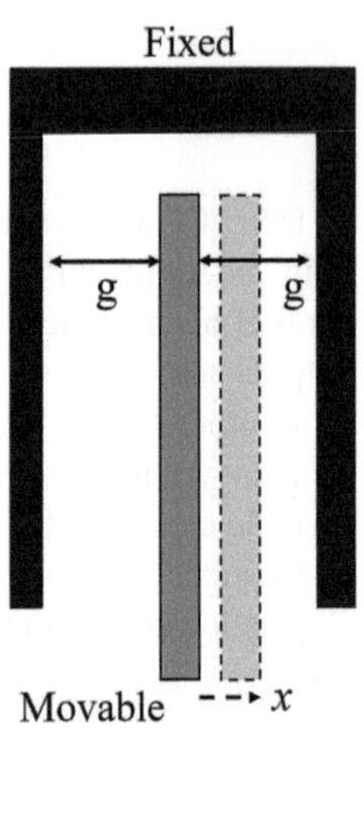

Fig. 6 Schematic of balanced capacitive structure unit cell based on a pair of equal gap parallel plate capacitors

As in the case of the any other capacitor topology, we can express the voltage-to-charge relation for the balanced capacitor structure as a Maclaurin's series as described previously by Eq. (12). In the case of the balanced capacitor structure, the coefficients of the Maclaurin's series are as follows,

$$G_1 = 0$$
$$G_2 = \frac{C_S}{(C_S + C_P)^2} \tag{18}$$

The coefficient of the DC term (i.e. G_0) is the same as the simple parallel plate capacitor topology as described in Eq. (18) as it corresponds to the case where the movable plate is at its equilibrium position (i.e. $x/g = 0$). As seen from Eq. (19), the coefficient of the term at the modulation frequency is zero. Given that the displacement of the central plate is harmonic, by applying the substitution relation in Eqs. (13)–(14) and (20), we obtain the following voltage-to-charge conversion relation,

$$V(t) \approx Q\left[\frac{1}{C_S + C_P} - \left(\frac{\bar{x}}{g}\right)^2 \frac{C_S}{2(C_S + C_P)^2} - \left(\frac{\bar{x}}{g}\right)^2 \frac{C_S}{2(C_S + C_P)^2}\cos(2\omega t)\right] \tag{19}$$

The above relation once again omits components beyond the second harmonic as only the components up to the second harmonic are of relevance to the analysis provided in this chapter. As the coefficient G_1 is zero, the amplitude of the component at the modulation frequency is thus also zero according to Eq. (14). The term of the interest is the second harmonic component. Equation (19) shows that the amplitude of the output voltage at the second harmonic is linear with the input charge. The ratio between the second harmonic output RMS voltage to input static charge can thus be used to define the responsivity of the variable capacitor electrometer,

$$S_R = \left(\frac{\bar{x}}{g}\right)^2 \frac{C_S}{2\sqrt{2}(C_S + C_P)^2} \tag{20}$$

The responsivity (denoted by S_R) of the electrometer is a figure of merit of interest for enhancing the resolution of the sensor. Figure 7 illustrates the relation between resolution, responsivity and offset. With reference to Fig. 7, the responsivity is defined by the slope of the linear calibration curve. As such, a higher responsivity is reflected in a steep slope. The offset is where the linear calibration curve meets the y-axis of the graph (i.e. the voltage when the input static charge is zero). Given that the level of input charge on the electrometer is inferred from the output voltage based on a calibration slope, the charge resolution of the electrometer is determined converting the output voltage noise to an equivalent charge noise in relation to the calibration slope,

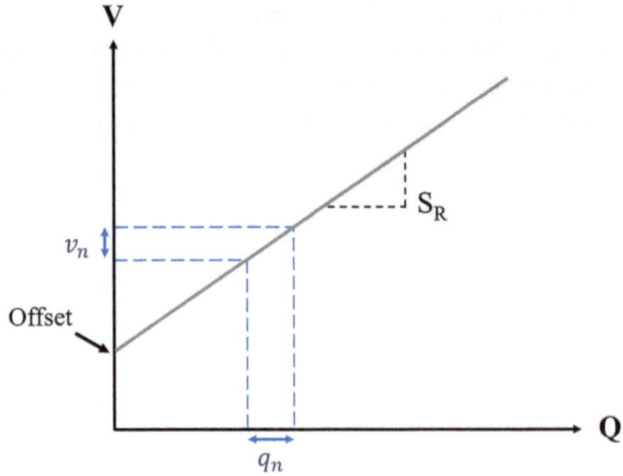

Fig. 7 Output RMS voltage (V) at twice the modulation frequency is a linear function of the static charge (Q). The slope of the best fit line defines the responsivity of the electrometer

$$q_n = v_n / S_R \tag{21}$$

For a given output voltage noise, a higher responsivity translates to a lower equivalent charge noise. As such, enhancing the electrometer's resolution must involve enhancing the responsivity of the variable capacitor, which will be considered in the following section. And for a given responsivity, a lower output voltage noise translates to a lower equivalent charge noise floor. As such, the input noise of the detection electronics must be considered for improving sensor resolution. Considerations of electronic noise will be discussed Sect. 7.

Figure 8 provides a schematic of full device comprising the variable capacitor at the centre of the entire structure, flanked by comb drives on each side oriented in opposite directions to produce linear differential drive. The balanced capacitor unit cell is arrayed to increase the total nominal sense capacitance (C_S). The central fingers of each unit cell are attached to the central shuttle which is driven into oscillation along the horizontal axis through the comb drives. The central shuttle is supported by four beam-type springs, one on each side of the shuttle, thus forming a flexural mode

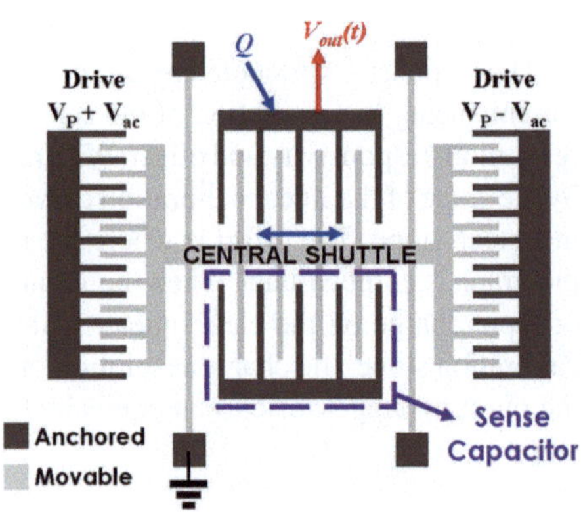

Fig. 8 Top view schematic of the time-modulated variable capacitor electrometer based on a balanced capacitor design. The variable capacitor is driven into oscillation through a pair of comb drives on each side of the shuttle

resonator. The spring constant of the resonator is determined by the sum of the spring constants of the springs. The mass of the resonator is the sum of the masses of the central shuttle (including the movable fingers of the variable capacitor) and movable comb drive fingers. The central fingers of the variable capacitor are grounded through the anchors of the springs, and input static charge is to be applied to the fixed fingers of the variable capacitor.

As can be seen from Eq. (20), the responsivity of the variable capacitor depends notably on the change in capacitor gap relative to its nominal value. As such, by operating the device at resonance, we can maximize the amount of displacement on the shuttle using the minimum possible level of AC drive. The degree of amplification in the response of the shuttle at resonance relative to off resonance operation scales proportionally with the quality factor. On this note, it is worth highlighting that having very high quality factors is not critical to improving the responsivity, though this may appear so at first glance. After all, higher quality factor provides for the same shuttle displacement with a lower electrostatic force. This means that the AC drive voltage amplitude can be reduced, consequently reducing charge feedthrough. While this is indeed correct, there are alternative ways to reduce AC drive apart from increasing quality factor. For example, the same output amplitude of electrostatic force can be generated with a lower AC drive voltage by increasing the DC bias applied to the comb drives. For reference, the variable capacitor electrometer reported in [3] was operated at ambient temperature and pressure, under which the quality factor was about

5 Effect of Parasitics on Sensitivity

From Eq. (20), we can see that the presence of the shunt parasitic capacitance (C_P) that appears parallel to the variable capacitor (C_S), lowers the responsivity for any given value of sense capacitance. The larger the value of C_P for a given value for C_S, the lower the resulting value of responsivity. In the hypothetical case where $C_P = 0$, then maximizing responsivity would simply require minimizing C_S as there would be no static capacitance component in the analysis. But given that some finite value of C_P will always exist, how should one design the variable capacitor to optimize the responsivity (S_R) for a given value of C_P? We see from Eq. (20) that the units for responsivity is the reciprocal of capacitance. As such, both the ratio between C_P and C_S as well their values determine S_R. To tie S_R to only the relationship of C_P and C_S, we here consider the dimensionless parameter $S_R C_P$. This dimensionless parameter can be expressed purely in terms of the ratio between C_P and C_S symbolised by the dimensionless variable of $\alpha = C_S/C_P$.

$$S_R C_P = \left(\frac{\bar{x}}{g}\right)^2 \frac{\alpha}{2\sqrt{2}(\alpha+1)^2} \quad (22)$$

In expressing the responsivity as a dimensionless objective function, $S_R C_P$, the value of this objective function is constant for a given value of α and a fixed amplitude of plate displacement. The result is a single curve showing the effect of varying C_S relative to C_P instead of having a family of curves, each corresponding to a certain value of C_P. Differentiating $S_R C_P$ in terms of α to find turning points, it can be shown that there is maximum point at $\alpha = 1$. In other words, for a given value of C_P, S_R is maximum when C_S is designed to be equal to C_P. This function is depicted by the graph shown in Fig. 9. It can be seen from the graph that when $\alpha < 1$ (i.e. when $C_S < C_P$), that the responsivity is much more sensitive to changes in the ratio of the two capacitances compared when $\alpha > 1$. From the perspective of design, when designing C_S to be as close as possible to the expected value of C_P, it is a more conservative choice to design C_S to be slightly larger than C_P rather than to be or slightly less. For reference, the graph in Fig. 9 is based on an amplitude of displacement that is a third of the gap. As S_R depends on $\left(\frac{\bar{x}}{g}\right)^2$, increasing the amplitude of displacement relative to the gap will increase S_R, though this contribution is negligible compared to reducing C_P and designing C_S to match C_P. More specifically, given that the optimal value for $S_R C_P$ is constant, this means that C_P must be reduced to increase S_R. A lower value of C_P would demand a lower value of C_S, which would mean a reduced number of fingers making up the variable capacitor. This would result in a smaller structure that would be less susceptible to acceleration effects as well as delivering a more compact device. Given that C_P plays a dominant role in determining S_R, it is important to identify the sources contributing C_P and thereafter to evaluate approaches to reduce C_P.

Sources contributing to C_P include parasitic shunt capacitances from the device structure, the interface between the device and the detection electronics, as well as the input impedance of the detection electronics. The origins of C_P from the device structure will be discussed in Sect. 6, with regards to the choice of fabrication process,

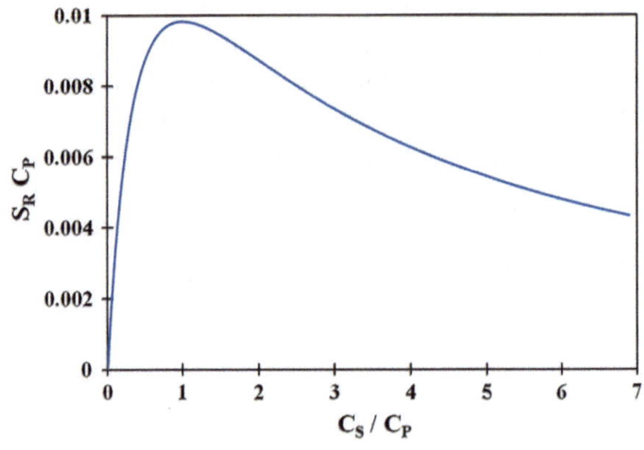

Fig. 9 Variation of the responsivity expressed a dimensional parameter as a function of the ratio of C_S to C_P, showing that $S_R C_P$ is maximum when C_S and C_P are equal

while the impact of the detection electronics on C_P will be discussed in Sect. 7. The issue of interfacial shunt capacitance between the device and the detection electronics will also be covered in Sect. 7.

6 Fabrication Process Considerations

The choice of fabrication process has an impact on the shunt parasitic capacitance (C_P), which has a notable effect on the responsivity of the variable capacitor electrometer. This section illustrates how the same type of fabrication process can have a significant impact on the shunt parasitic capacitance intrinsic to the device. The example given here is based on a conventional silicon-on-insulator (SOI) micromachining processes, where the key difference lies in the approach of how the device is released to create a freestanding movable structure. In one case, the device is released from the top side of the movable structure while in the other case, the device is released from the back side of the movable structure. Apart from the release step, the other steps of the process (that all precede the release step) are identical.

The steps preceding the release step of the SOI micromachining process are illustrated in Fig. 10.

The SOI micromachining process begins with a highly doped n-type silicon device layer. The silicon device layer of SOI wafer forms the structure layer that defines the variable capacitor, comb drives and springs. The high level of doping is necessary to create an ohmic contact when fabricating the contact pads of the silicon device

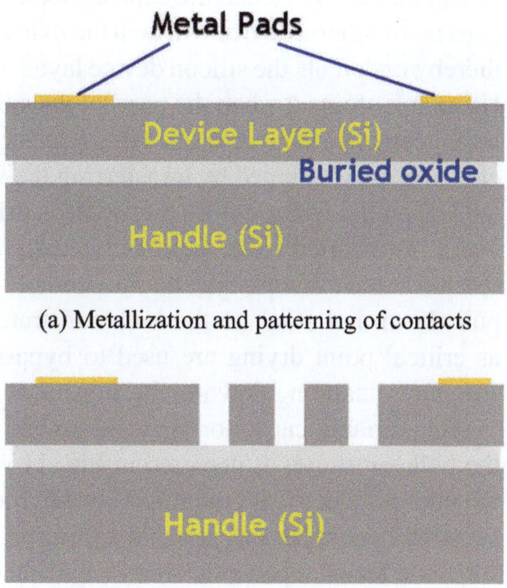

(a) Metallization and patterning of contacts

(b) Deep reactive ion etching of silicon device layer

Fig. 10 Fabrication process steps of the silicon micromachining process prior to the final release step layer. Metal contacts are deposited and patterned on the device layer. For the metal contacts, a thin layer of Cr (e.g. 5 nm) is deposited first to serve as an adhesion layer followed by depositing a metal with low resistivity such as Au to minimize the electrical losses. On the choice of metal for the contacts, it is crucial to select a metal that is resistant to etching by hydrofluoric (HF) acid, the etchant used subsequently in the release step to etch the buried oxide layer. This step of metallization is shown in Fig. 10a. Next, the various features of the silicon structure that defines the variable capacitor, both anchored and movable parts, are simultaneously patterned by Deep Reactive Ion Etching (DRIE) from the front side. Using DRIE to pattern the various features in the silicon device layer allows the creation of deep and vertical side walls. This capability to create deep and vertical side walls is essential to defining the capacitor plates that make up the comb drives and variable capacitor. The result of the DRIE step is illustrated in Fig. 10b. The deep etching process stops at the buried oxide layer which provides an etch stop. After the silicon structure has been defined using DRIE, the next step is to release the structure, either from the top side or from the back side.

6.1 SOI Top Side Release

In the top side release approach, the buried oxide layer beneath the silicon device layer is etched through the openings in the silicon device layer where silicon was previously away by DRIE. The buried oxide layer could be etched by hydrofluoric acid (HF) either in the vapour phase or as a wet etchant. The etching process is isotropic. Having etched the exposed oxide in the gaps, the HF entering the side wall gaps in the silicon device will etch the oxide sideways. The sideways etch of the oxide thereby undercuts the silicon device layer. The etching process is timed, whereby the HF etch is stopped when the parts of the silicon device layer that should be undercut have been released from bulk silicon substrate. In the case of wet etching by HF, the etch process is stopped by transferring the wafer to water bath to successively dilute the HF on the wafer before the wafer is dried. Care must be taken when drying the wafer. The buried oxide layer is typically around 1 μm. Due to the strong surface tension at the micron scale, the suspended structures in the silicon device layer could pull down and stuck to the bulk substrate as the wafer is left to dry. Steps such as critical point drying are used to bypass the liquid phase to avoid the released structures snapping down to the substrate.

The variable capacitor is found among the released structures, suspended over the bulk substrate, as depicted in Fig. 11. The bulk substrate is typically grounded to suppress electrical coupling between the drive and sense ports through the bulk substrate. As such, the variable capacitor fingers suspended over the ground bulk substrate forms a parasitic shunt capacitance between sensing node of the variable capacitor and ground. The larger the size of the array of fingers forming the variable, the larger will be the shunt parasitic capacitance to ground. In this case, C_P is thus intrinsic to the structure. Therefore, even without connecting to the interface

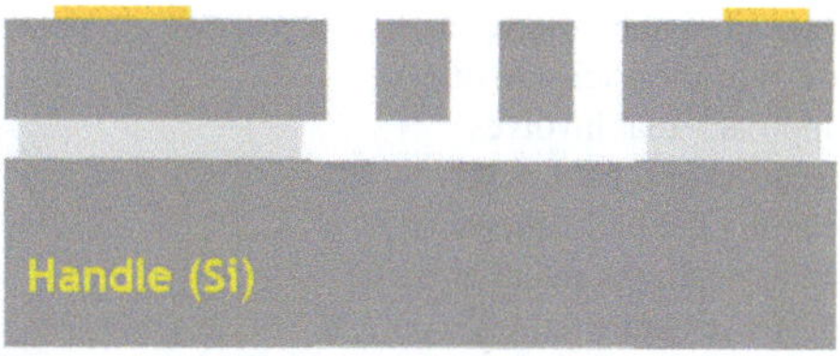

Fig. 11 Cross-section view of micromachined variable capacitor electrometer fabricated by releasing the structure from the top side by etching away buried oxide layer with HF

detection electronics that will further increase C_P, C_P is already introduced in the structure itself. As mentioned previously in Sect. 5, a larger value of C_P lowers the responsivity of the variable capacitor electrometer, and thus C_P should be minimized where practically possible. In the case of the front side release approach, the presence of the underlying bulk substrate adds to C_P.

6.2 SOI Bottom Side Release

In the bottom side release approach, the etching of the buried oxide layer is carried out from the back side of the wafer. However, as illustrated by Fig. 10b, for the HF to access the buried oxide from the back of the wafer, the underlying bulk silicon layer must be removed first. A long etch is required to remove the bulk substrate due the thickness of the layer (>400 μm). Prior to the etch of the bulk substrate, the structures fabricated on the front side of the wafer must be protected first. This could be done by covering the front side with protective polyimide coat temporarily (to be removed subsequently), as depicted in Fig. 12a. The wafer can then be rested on the front size to process the wafer on the back side. Processing on the backside of the wafer begins with etching through the bulk silicon substrate to create a deep trench with close to vertical side walls using DRIE. Once again, the buried oxide layer provides a etch stop. Once the cavity through the bulk silicon substrate has been etched, the buried oxide layer is exposed from the backside as illustrated in Fig. 12b. This is followed by etching the buried oxide layer from the backside using HF. Unlike the front side release approach, there is no sideways etch required to undercut the silicon device layer as the buried oxide beneath the silicon device layer structures to be released are directly exposed to HF. Figure 12c shows the resulting structure after the buried oxide has been removed to expose the bottom side of the silicon device layer in the structures to be released. Finally, to release the structure after all the etching steps have been completed, the protective polyimide coat on the front side of the device layer is removed. The final released structure is depicted in Fig. 12d.

Comparing these two approaches of releasing the structures, the backside release approach is more time consuming and involves more processing steps. However, this cost of additional processing complexity and duration comes with the benefit of greatly lowering the parasitic shunt capacitance inherent to the device. As discussed previously in Sect. 5, lowering the overall parasitic shunt capacitance as the sensing node of the variable capacitor electrometer is crucial to enhancing responsivity.

Fig. 12 Fabrication process steps of the backside release approach that involves creation of a back cavity by etching the bulk silicon layer underneath the released parts of the structure

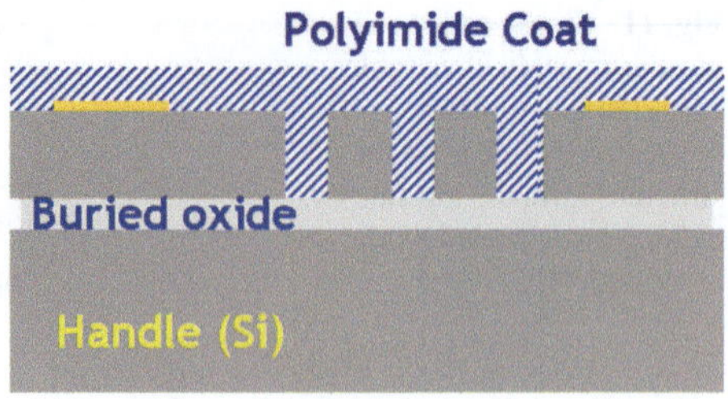

(a) Protection of device layer on front side with polyimide coat

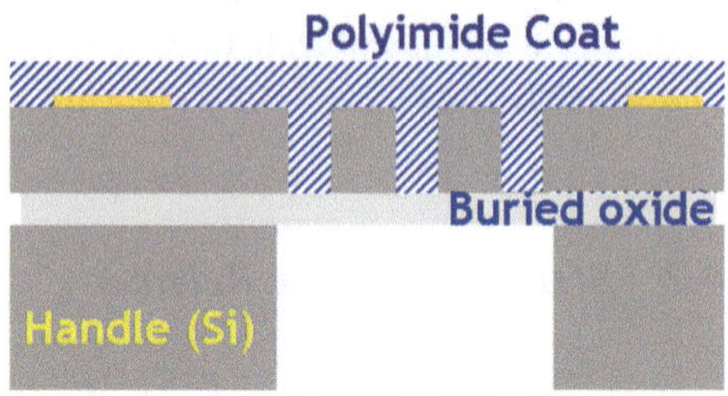

(b) Deep reactive ion etching of bulk silicon handle layer

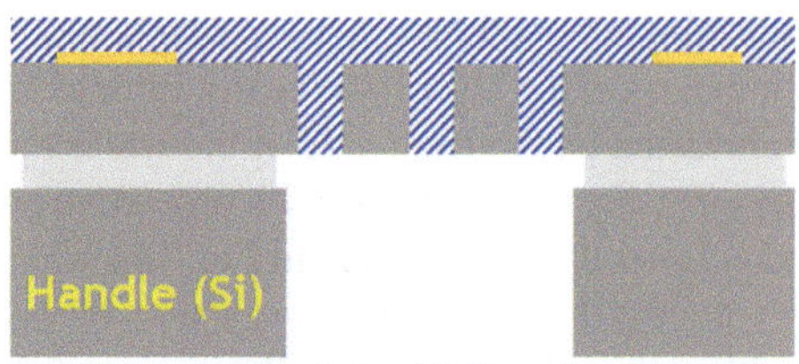

(c) HF etch of buried oxide layer from the back cavity

(d) Removal of polyimide protective coat

In the case of the backside release approach, the bulk substrate under the released structures in the silicon device layer is removed. As such, the shunt parasitic capacitance between the variable capacitor fingers and the grounded bulk substrate that is present in the frontside release configuration, is absent in the backside release configuration. Hence compared to the frontside release configuration, the backside release configuration suffers from much lower shunt parasitic capacitance intrinsic to the device.

7 Design of Interface Electronics

As mentioned in Sect. 5, the resolution of the electrometer is determined by both the responsivity of variable capacitor as well as the noise of the interface electronics for reading out and amplifier the voltage converted from the static charge input to be sensed. The noise limit of variable capacitor electrometers is generally set by electronic noise over thermomechanical noise. While thermomechanical noise is known to dominate forced-based sensors typically, their contributions to variable capacitor electrometers is insignificant. Using variable capacitor electrometers reported in the literature as reference, the Brownian motion is around six orders of magnitude lower than the amplitudes of the moving showing shuttle and capacitor plates. Referring to Eq. (20), we see that the responsivity scales with the square of the plate displacement. As such, where the Brownian motion is 1 ppm of the expected actuated displacement from the comb drives, the associated noise in the responsivity (i.e. output of the variable capacitor per unit charge) will be twelve orders of magnitude lower than nominal value of responsivity. Therefore, compared to the noise from the interface electronics, the contribution from thermomechanical noise is insignificant.

With respect to electronic noise, of most interest in this chapter is flicker noise, which has been mentioned at the beginning of this chapter. All active devices exhibit flicker noise, which arises from the random capture and release of carriers due to the presence of carrier traps associated with crystal defects and contamination. The result is a statistical variation concentrated at lower frequencies. For field-effect transistors (FET), the flicker noise spectrum density for a known direct current (I_D) is modelled by a 1/*f* function, hence the term 1/*f* noise,

$$\frac{i_n^2}{\Delta f} = \beta \frac{I_D^\alpha}{f} \tag{23}$$

where i_n is the current noise, Δf is the bandwidth, β is a constant that depends on the type of device and fabrication technology. The exponent α varies between 0.5 and 2. As mentioned previously, the frequency at which the 1/*f* slope coincides with the white noise is called the noise corner or corner frequency. The value of the corner frequency depends on the type of active device as quantified by the parameter β, which differs from one type of device to another. For instance, the corner frequency of a metal oxide semiconductor (MOS) FET can be rather high (up to MHz range) compared to a JFET for which the noise corner can be lower than 1 kHz.

7.1 Junction Field Effect Transistor Buffer

Due to its comparatively low corner frequency, the JFET is an attractive device to interface with the sense node of the variable capacitor. A lower $1/f$ noise corner frequency allows for greater flexibility in design the device. As mentioned in Sect. 1, the motivation for using a variable capacitor as an electrometer is to up-modulate and convert a static charge input to an AC voltage with a frequency component that is higher than the $1/f$ noise corner of the interface circuit. As such, having a lower $1/f$ noise corner reducing the demand to design resonators that would otherwise require higher resonant frequencies. Given the practical limits in reducing C_P and considering that C_S should match C_P to optimize the responsivity, it is difficult to significantly reduce the mass of the variable capacitor structure. Furthermore, the springs cannot be too stiff in order to minimum the required drive voltage to actuate the variable capacitor structure into oscillation. As such, it is challenging to design structures to have resonant frequencies at the MHz range if MOSFETs were used instead of JFETs. The other advantage of employing a JFET as the interface device, as with FET devices in general, is the high input impedance that is typically required for charge detection. The JFET also has a relatively low gate leakage current (pA level). A low leakage current is important to minimize the rate of discharge to the shunt capacitors. Figure 13 shows how a high input impedance and unity gain JFET buffer source can implemented with two JFETs. Q2 and R_S together form a current source that can be tuned through adjusting R_S. The current source is connected to the source of Q1, the gate of which is connected to the sensor node of the variable capacitor. The output of the buffer is taken from the source of Q1, which can then be amplified in the second stage of an amplifier.

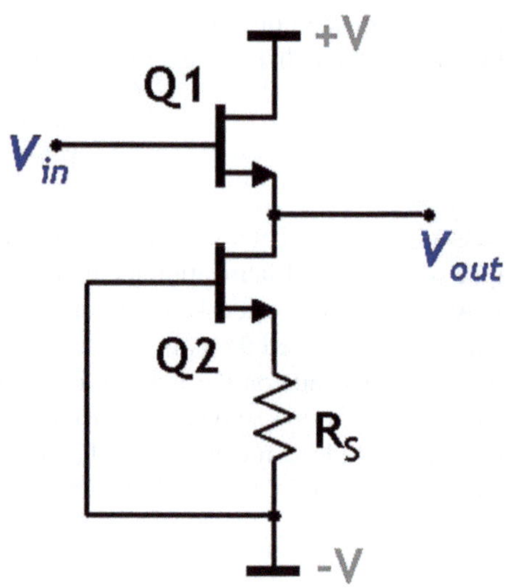

Fig. 13 Circuit schematic of a junction field effect transistor (JFET) buffer

7.2 Trans-Capacitance Amplifier

While the leakage gate current of a JFET is much lower compared to other active devices, its effect on the detected AC voltage from a variable capacitor is still clearly observable in the form of a prominent offset signal. In the case of an n-channel JFET where the leakage current charges up the shunt capacitance, inputting negative charges on the variable capacitor results in temporary fall in the offset, which after some time rises back up. One alternative circuit topology to address the issue of input current leakage is the trans-capacitance amplifier, which is depicted in Fig. 14. In a trans-capacitance, a capacitor is placed between in the feedback connection (C_2), while an input capacitor is placed between the sense node of the variable capacitor and the inverting input of the operational amplifier. Hence even though the input transistors of the operational amplifier may sink or source a current, this bias current is isolated from the variable capacitor by the capacitor C_1 acting as a direct current block. A feedback resistor (R_F) is included in the feedback connection to provide biasing to the transistors in the operational amplifier. The impedance of R_F must be much larger than that of C_2 at the frequency of detection so that R_F appears as an open circuit relative to C_2. In this configuration, the input current C_1 defines the input impedance of the circuit. As such, the load seen by the variable capacitor is the capacitor C_1. Therefore, C_1 appears as an additional shunt capacitance contributing to C_P, such that now,

$$C_P' = C_P + C_1 \tag{24}$$

The gain of the trans-capacitance amplifier is defined by the ratio of impedances between the feedback branch and input branch, such that,

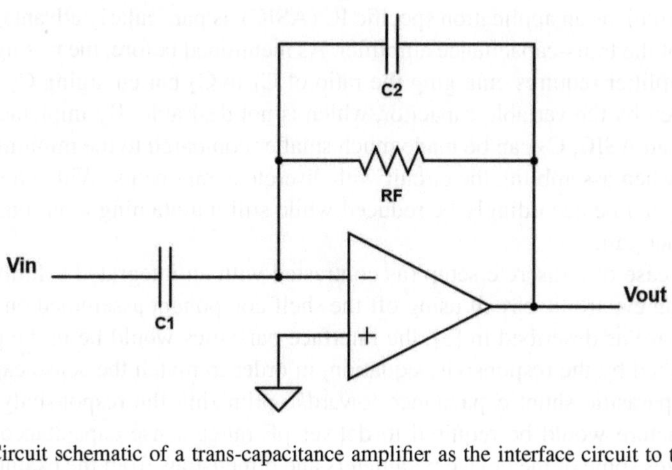

Fig. 14 Circuit schematic of a trans-capacitance amplifier as the interface circuit to the MEMS variable capacitor

$$\frac{V_{out}}{V_{in}} = \frac{C_1}{C_2} \tag{25}$$

Therefore, for higher gain, C_1 should be as large as possible while C_2 should be as small as possible. But a higher C_1 translates to a higher C_P and thus lower responsivity. As such, the enhancing the gain of the amplifier comes at the cost of lowering the responsivity. However, the trans-capacitance amplifier interface circuit has the benefit of greatly reducing the charging effect at the input due to leakage current.

7.3 MEMS and Readout Electronics Integration

Given that C_P degrades the variable capacitor electrometer's responsivity, reducing the parasitics at the interface between the variable capacitor and detection circuit can further improve the responsivity and resolution of the electrometer. In this regard, interface parasitics can be practically reduced through tighter integration between the micromechanical device and the interface circuit. One approach of integration is monolithic integration whereby the micromechanical device and interface circuit are fabricated on the same substrate. This approach non-trivial process development such that fabrication of each of the parts are compatible. The other approach is a two-chip solution where the two parts are fabricated separately and stacked together at the end. Just as in the monolithic integration approach, the readout electronics are implemented as an integrated circuit (IC) platform but fabricated separately from the micromachining process used for the micromechanical device. The connections between the two chips, which greatly determines the interface parasitics, could be done either using through silicon vias (TSVs) or flip chip bonding. Implementing the readout circuit as an application specific IC (ASIC), is particularly advantageous for the case of the trans-capacitance amplifier. As mentioned before, the raising the gain of the amplifier requires enlarging the ratio of C_1 to C_2 but enlarging C_1 increases the C_P seen by the variable capacitor, which is not desirable. By implementing the circuit as an ASIC, C_2 can be made much smaller compared to the minimum values possible when assembling the circuit with discrete components. With a lower value of C_2, C_1 can be accordingly be reduced while still maintaining a reasonable level of amplifier gain.

In the case of a discrete setup (as contrasted with an integrated solution) where the sensing electronic circuit using off the shelf component assembled on a printed circuit board as described in [3], the interface parasitics would be in the pF range. As described by the responsivity equation, in order to match the sense capacitance with the parasitic shunt capacitance towards optimizing the responsivity, a rather large structure would be required to deliver pF range sense capacitance. Table 1 summarizes some of the device parameters and dimensions from the example in [3].

Table 1 Summary of device parameters and dimensions from the example reported in [3] based on a backside release SOI micromachining process

Device parameter or feature	Value (unit)
Structure thickness (height of capacitance and comb drives)	25 (μm)
Sense capacitor gap (i.e. variable capacitor nominal gap)	5 (μm)
Number of sense capacitor finger pairs per side for the variable capacitor	145
Sense capacitance of the variable capacitor	1.28 (pF)
Number of comb drive fingers per side	684
Spring constant of mechanical springs	33 (N/m)
Mass of shuttle with attached comb drive fingers and variable capacitor fingers	100 (μg)
Resonant frequency	2.9 (kHz)
Total length of device including comb drives, sense capacitors and shuttle (approximate)	3.2 (mm)
Total width of device including comb drives, sense capacitors and shuttle (approximate)	2 (mm)
Deduced value of shunt parasitic capacitance	2 (pF)

8 Temperature Sensitivity

The Young's modulus of silicon is known to have a temperature coefficient of around 60 ppm/K typically. Therefore, in the case of a resonator, fluctuations in the ambient temperature would cause the resonant frequency of the device to also fluctuate accordingly. It is worth recalling that the variable capacitor is driven at its resonant frequency to maximize the amplitude of displacement for a given drive voltage. When the resonant frequency of the structure changes according to the ambient temperature, and if the driving frequency is fixed and thus not adjusted to match the new resonant frequency, the displacement amplitude of the structure will be reduced. This leads to a fall in the responsivity of the electrometer. As such, if the drive frequency is not actively tuned to match the changing resonant frequency, then drifts in the ambient temperature will lead to drifts in the responsivity of the electrometer. The effect of temperature sensitivity on responsivity becomes even more prominent when the quality factor is higher. As such, to stabilize the responsivity over large temperature changes, the oscillation frequency of the variable capacitor must be automatically tuned to the natural frequency of the structure as it varies over temperature. One way to ensure that the oscillation frequency of the variable capacitor is locked to the natural frequency of the structure is to implement a self-oscillation closed loop. In the open loop configuration, the structure is driven at a fixed frequency tuned initially to match the natural frequency of the structure. But the drive frequency is unable to respond to changes in natural frequency due to changes in the ambient temperature. To track the natural frequency of the structure, we need to sense the oscillations apart from the output of the variable sense capacitor for converting static charge to voltage. This requires modification of the comb drive layout attached to the shuttle.

Instead of having only two sets of comb drives (used only for differential actuation), we include an additional two sets of comb drives to differential sense the shuttle's oscillation by capacitive sensing. As the fingers of the comb drives displace relative to each other, the charge stored on the comb drives is modulated to yield a motional current (i_{out}) at the drive frequency,

$$i_{out} = \omega V_{dc}\left(\frac{dC}{dx}\right)x \tag{26}$$

The modified structure configured for self-oscillation is depicted in Fig. 15. As shown in Fig. 15, the motional current ($\mathrm{i_{out}}$) from the sensing comb drives to fed to a trans-resistance amplifier. To realize self-oscillation, the overall circuit should meet the Barkhausen criteria, which requires that the open loop gain is greater than unity

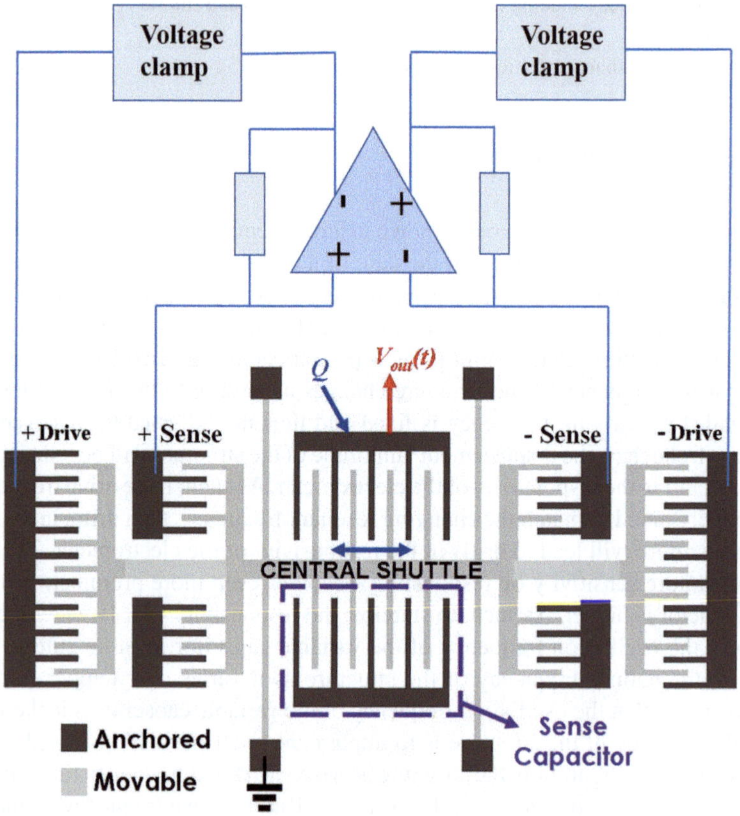

Fig. 15 Top view schematic of the modified comb drive layout of the time-modulated variable capacitor electrometer configured to allow self-oscillation. The variable capacitor is driven into oscillation through two sets of comb drives on each side of the shuttle. The oscillations are sensed through another two sets of comb drives on side of the shuttle

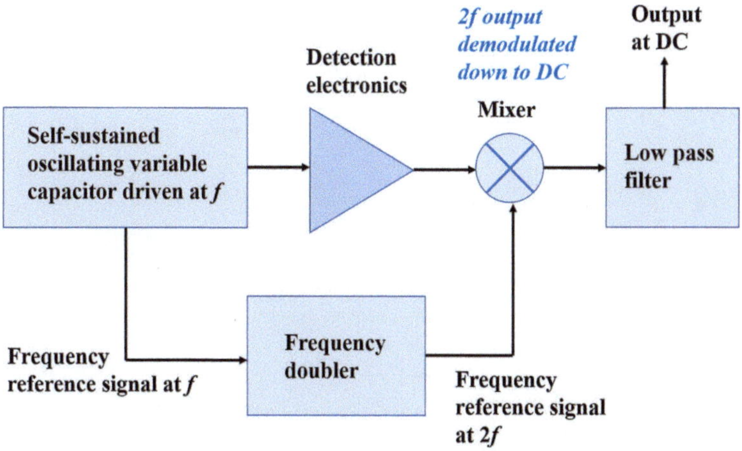

Fig. 16 System-level diagram of phase lock loop circuit for detection of the sensed voltage from the variable capacitor at the second harmonic of the drive frequency

and a phase shift of 0°. A voltage clamp is included in the amplifier circuit to hard limit the drive voltage to ensure that the shuttle is driven at the same voltage level, as the oscillation amplitude is proportional to the AC drive voltage level. Maintaining the same drive voltage level keeps the oscillation amplitude of the shuttle constant to ensure that the responsivity does not fluctuate due to variations in the oscillation amplitude.

To implement a self-contained sensor system without measuring instruments, one could use a phase lock loop (PLL) to detect the output at the second harmonic. One implementation of the PLL circuit is given in Fig. 16. In the PLL circuit shown, one of the two AC voltage drive inputs is then fed to a buffer. To monitor the sensed voltage at the variable capacitor sense node at twice the drive frequency, the output of the buffer is used as a frequency reference. This frequency reference at the drive frequency is then put through a frequency doubler to produce a second harmonic frequency reference. Through a mixer followed by a low pass filter, the second harmonic reference signal demodulates the second harmonic sensed voltage output from the interface detection electronics down to DC. The low pass filter attenuates the higher harmonic components.

References

1. R. Gunn, Principles of a new portable electrometer. Phys. Rev. **40**, 307–312 (1932)
2. P.S. Riehl, K.L. Scott, R.S. Muller, R.T. Howe, J.A. Yasaitis, Electrostatic charge and field sensors based on micromechanical resonators. J. Microelectromech. Syst. **12**(5), 557–589 (2003)
3. J.E.-Y. Lee, Y. Zhu, A.A. Seshia, Room temperature electrometry with SUB-10 electron charge resolution. J. Micromech. Microeng. **18**, 025033 (2008)

Micromachined Vibrating-Reed Electrometer in Silicon-on-Glass Technology

Yong Ruan, Jubayer Jalil and Yong Zhu

Abstract Designing charge sensors for electrometry is deemed significant because of the sensitivity and resolution issues in the range of micro-scales. This chapter presents the design of microelectromechanical systems (MEMS) vibrating-reed (VR) electrometer developed based on micromechanical variable capacitors in silicon-on-glass (SOG) process. By using vibrating-reed technique, the resonator is driven at the frequency above the corner frequency of Flicker noise and the charge measurement is performed at the second harmonic of the resonator's frequency. This chapter also describes the noise characterization and the referred-to-input (RTI) noise reduction methods for operational amplifier (opamp) based preamplifier. The design of an improved noninverting topology overcame the issues of leakage current and charge resetting during measurement. A comprehensive noise model has been proposed considering all electronic noise contributors including the preamp in conjunction with capacitive MEMS sensor in silicon-on glass (SOG) process and ancillary components. The research findings proved that the blocking capacitor and shunt resistor have significant impact on the RTI noise of the preamp and the charge resolution of the electrometer. The noise analysis and measurement results provide a practical guideline for low noise electrometer interface circuit design. Trade-offs among opamp selection, blocking capacitor value, shunt resistor value etc. have been discussed to provide an original and in-depth analysis for noise and resolution performance optimization in the design of MEMS electrometers.

1 Introduction

Nowadays, in big urban cities and heavy industrial areas, aerosol particles or airborne particulate matters cause a severe impact on human health owing to rampant air

---

Y. Ruan
Department of Precision Instruments, Tsinghua University, Beijing, People's Republic of China
e-mail: ruanyong@mail.tsinghua.edu.cn

J. Jalil · Y. Zhu (✉)
School of Engineering and Built Environment, Griffith University, Gold Coast Campus, Gold Coast, QLD 4222, Australia
e-mail: y.zhu@griffith.edu.au

Y. Zhu (ed.), *Micro and Nano Machined Electrometers*,
https://doi.org/10.1007/978-981-13-3247-0_5

pollution. To accurately detect air quality in the environment, aerosol electrometry is one of the promising techniques to measure nano- and micro-sized aerosols in our daily life. One of the advantages of this measuring instrumentation is to provide real-time readings of particle concentration as a total count or a distribution of particle size. Electrometry plays a vital role designing such an apparatus. This is because a direct-contact charge electrometer or charge sensor is built and integrated into this apparatus for electrostatic charge measurement. In the first decade of the 20th century, a concerted effort was carried out to determine the electrostatic charge on the electron. In 1910, A. J. Millikan's fabled nonvolatile oil drop experiment resulted in the determination of the charge on the electron, which led to his Nobel Prize for Physics in 1923 [1]. His investigation has had a major impact on the study of single micro-particles. Currently, electrometers have a broad range of applications besides detection of ultrafine particles, such as those involving the measurement of ionization radiation, detection of bio-analyte, mass spectrometry, scanning tunneling microscopy, and quantum computation.

Generally, heterogeneity and substrate differences in sensor manufacturing, excessive inherent noises from sensing elements, electronic circuits and external disturbances are a few of the reasons that limit a micro-charge sensor's output accuracy, which in turn hinders its usage in the applications where detection of weak charge signals is required. Presently, a widely used commercial electrometer, for example, Keysight B2987A demonstrates poor resolution of 6,300 electrons at room temperature and ambient pressure. Conversely, nano-capacitive gate sensing single electron transistors (SETs) [2, 3], and nanoelectromechanical systems (NEMS) resonator-based electrometers [4] can detect a fraction of electrons under stringent conditions of cryogenic temperature and vacuum pressure. Moreover, these devices are fabricated using non-standard process technologies. To overcome these issues, micro-electromechanical systems (MEMS) based electrometers, such as resonator type and vibrating-reed (VR) type are suitable alternatives due to their matured fabrication technologies and practical operation under room temperature. However, MEMS resonator type electrometers [5, 6] require vacuum chambers to achieve a high quality factor (Q) to exhibit a better charge resolution. In contrast to resonant electrometers, micromachined VR electrometers [7–9] can achieve comparably a better charge resolution at room temperature without vacuum conditions.

2 MEMS Vibrating-Reed Electrometry

One of the reasons for poor resolution of commercial electrometers is the use of DC preamplifiers (preamps) with a high magnitude of Flicker noise or $1/f$ noise (at near DC or low frequencies). To stave off the Flicker noise, the MEMS VR electrometer, mainly comprising of a vibrating capacitor with a driving circuit and an AC preamp, modulates a static weak charge signal to a frequency higher than the corner frequency of Flicker noise. In these electrometers, a tiny amount of charge (or electrons) is deposited on the immovable electrode of the parallel-plates of the

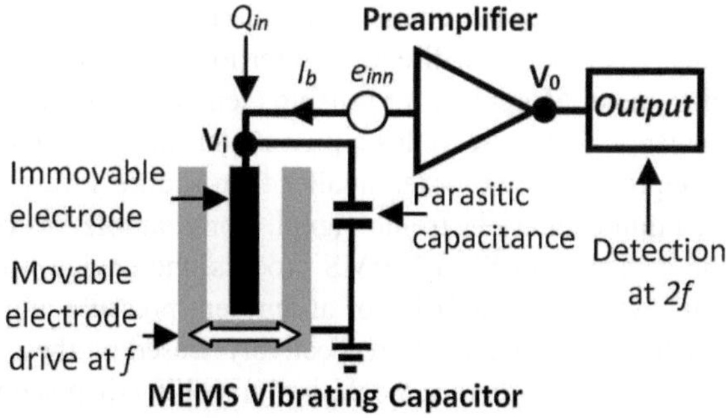

Fig. 1 Conceptual schematic diagram of MEMS vibrating-reed (VR) electrometer [10]. Copyright 2019. Reproduced with permission from IEEE Transactions on Instrumentation and Measurement

vibrating capacitor, as shown in Fig. 1. The charge is then converted to an AC voltage V_i due to the mechanically movable electrode, which is vibrated by an external force having a driving frequency (f) in a cyclic manner. Eventually, the AC signal from the output voltage V_0 of preamp is detected at the second harmonic frequency ($2f$) using a signal measurement equipment.

The charge resolution of the electrometer is quantified as

$$Q_n = e_{inn}/S_e \tag{1}$$

where S_e denotes the charge-to-voltage (rms) conversion gain or sensitivity of the MEMS vibrating capacitor, and e_{inn} denotes the total equivalent referred-to-input (RTI) noise voltage. Q_n represents the minimum input charge required at the input of electrometer to produce an output signal-to-noise ratio of 1. It is perceived from Eq. (1) that Q_n is the function of two attributes: the sensitivity and the total equivalent RTI noise voltage at the frequency of interest. Thus, during charge electrometry at room temperature, both higher the sensitivity of the sensing device and lower noise floor at the input of the preamp stage are of importance to achieve higher resolution of the electrometer.

Both electronic noise and mechanical noise can have significant impact on the improvement of resolution for many MEMS sensors. For MEMS VR electrometers, the moving plates of the vibrating capacitor is a mechanical resonator with a micro-meter scale traveling range yielding a relatively small electronic signal during measurement. We know that thermo-mechanical motion of the MEMS device may set a fundamental limit for the intrinsic noise of the system. This noise is known as Brownian noise. The Brownian noise or thermo-mechanical displacement power spectral density (per unit bandwidth) is expressed as

$$S_X = \frac{4k_B T Q}{m\omega_0^3} \tag{2}$$

where m is the mass of the moving shuttle, k_B is Boltzmann's constant, T is the operating temperature, Q is the quality factor of resonator, and ω_0 is the angular frequency of the moving shuttle.

In this case, the charge resolution is determined by the random displacement noise (S_X), which is directly proportional to Q. Thus, lower Q leads to a smaller displacement. Moreover, for sensing element in kHz range, the damping coefficient of the Brownian noise yields a pico-meter range Brownian noise displacement, which is approximately 6-order smaller than shuttle's travelling range and corresponds to an infinitesimal electronic signal. For example, according to the designed vibrating capacitor in SOG-MEMS process, the device was vibrated at 2.8 kHz, and its effective mass and Q factor at ambient pressure and room temperature were estimated as 105 nkg and 8, respectively. Based on these values, the Brownian noise is 0.48 pm/rtHz, whereas the desired travelling displacement of the vibrating capacitor was approximately 2.8 μm, which was 5.8×10^6 times higher than the displacement due to the Brownian noise. Hence, Brownian noise has an insignificant amount of contribution to the charge resolution at the output in the vibrating capacitor.

The main obstacle in the realization of the high-resolution VR charge sensors is their intrinsic electronic noise sources from the amplifying devices of the preamp, especially at the frequencies below the corner frequency of Flicker noise (1/*f* noise). Flicker noise is a nuisance to the circuits at very low frequencies. For MOSFETs, there are three different theories to explain the physical mechanisms of Flicker noise [11]: the McWhorter number fluctuation theory model, the Hooge mobility fluctuation theory model and the unified 1/*f* noise theory model. However, the origin of this noise is still an unresolved problem in semiconductor physics. This is because there is still no conventional accepted theory on 1/*f* noise. At a higher frequency range above the corner frequency of 1/*f* noise, both Johnson (thermal) and shot noises become dominant. Thus, minimizing the noise of the front-end electronics above the corner frequency of 1/*f* noise is decisive in obtaining a high-resolution of charge sensors.

Apart from the intrinsic input noises of the amplifying devices, the overall noise of the preamp for MEMS VR charge sensors depends on a number of crucial factors, such as circuit topology, process technology of the semiconductor devices, biasing and gain resistors of the circuit, input capacitance and input resistance of the amplifying device, operating frequency of the vibrating sensor, and sensing element's capacitance and parasitic capacitance due to fabrication technology. To develop an optimized preamp, the circuit designers must control over all of the above factors, except the mechanical operating frequency f because it is determined by the MEMS sensor itself. For low operating frequencies less than 10 kHz, the impedance of the input capacitance of the preamp cannot be neglected. Due to the small amplitude of periodic voltage signals from a MEMS VR sensor, amplification of the input signal must be performed before the output signal can be processed. In addition, the DC input bias or leakage current I_b is another paramount issue that needs to be considered (as shown in Fig. 1). The magnitude of the leakage current is required to be as low as possible for the overall efficacy of the readout circuit.

In the last 15 years, researchers have implemented various preamps for MEMS VR electrometer using different approaches. Riehl et al. [7] developed an on-chip CMOS preamp in silicon-on-insulator (SOI)-MEMS process, which integrated buffer circuit, gain amplifier and MEMS switch to reset charge. Lee et al. [12, 13] designed a hybrid MEMS VR electrometer in SOI-MEMS process, where interface circuit was implemented by an off-the-shelf JFET without resetting gate charge arrangement. None of the above circuits were taken any steps to hinder unsolicited input charge from the preamp to the MEMS sensor. Zhu et al. [14] constructed an SOI-MEMS electrometer using an inverting-type operational amplifier (opamp) based preamp with a charge blocking capacitor. To achieve high gain and charge resolution, the feedback capacitor's value must be small, for example less than 1 pF. However, it is not practical to obtain an accurate sub-pF capacitance in standard PCB implementation due to stray capacitance of the PCBs. Unlike Zhu's capacitor-based preamp, Jaramillo et al. [15] built an SOI-MEMS electrometer using non-inverting resistor-based preamp. The circuit has enough gain to amplify low input voltage signal and the utilization of an electronic switch can reset input charge. However, the electrometer is prone to charge drift due to the leakage current in the absence of a charge blocking capacitor. The input leakage current in the opamp-based preamp creates two main consequences: shot noise and offset voltage from the external high input impedance. The state-of-the-art opamps have been used in designing the preamp circuits in numerous sensing applications because of their continual improvement of noise characteristics. Moreover, chip manufacturers provide adequate information on opamps which are helpful for circuit designers to characterize those devices accurately. Therefore, to comply with the aforementioned design requirements, a commercial off-the-shelf opamp-based preamp is portrayed in this work for a hybrid MEMS vibrating-reed charge sensor.

3 Silicon-on-Glass Technology

Until today, most micromachined charge sensing devices have been fabricated on silicon-on-insulator (SOI) substrates. Nonetheless, the existing SOI-MEMS vibrating-reed based electrometers undergo difficulties in achieving high sensitivity due to their inherent parasitic capacitance (C_p), which is comparable to the time-modulated sense capacitance (C_{var}) in the order of pF. In SOI-MEMS process, a single crystalline silicon (SCS) device layer lies on a buried oxide layer setting on top of a silicon substrate as shown in Fig. 2. Therefore, in SOI-MEMS devices, a major contributor to the parasitic capacitance is the static capacitance (C_{p1}) between the sense electrodes and grounded substrate. Even though the underlying substrate of the device is removed, the parasitic capacitance still exists in the pad area because of silicon-oxide-silicon sandwich structures, as shown in Fig. 2.

To overcome the inherent issue of the static parasitic capacitance between the pad and grounded substrate in SOI-MEMS process, a unique silicon-on-glass (SOG) based vibrating-reed charge sensing system has been proposed and is presented in this chapter. The SOG fabrication process shown in Fig. 3 is implemented with the advent of anodic bonding between silicon and glass wafers. This wafer bonding technique allows two wafers of different materials (silicon and glass) to be bonded together using a sufficiently powerful electrostatic field. Anodic bonding process

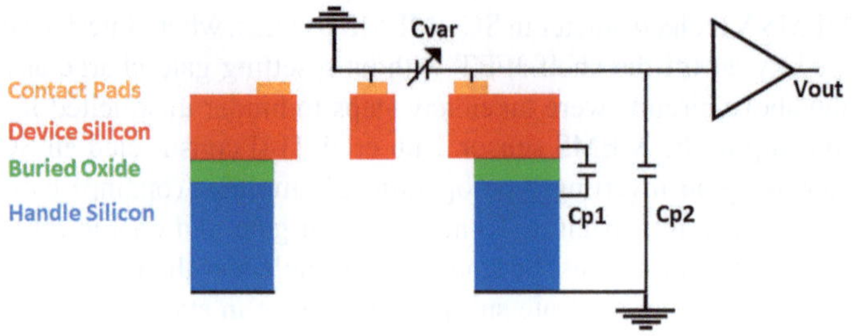

Fig. 2 Cross-sectional view of silicon-on-insulator (SOI) MEMS electrometer layers with capacitance sources labelled

can form a strong atomic Si-O-Si chemical bond between silicon and sodium-rich Pyrex glass wafers. Since silicon and Pyrex glass have similar thermal expansion coefficients, they have been widely adopted as suitable materials to be anodically bonded as substrates for MEMS devices.

As a bulk micromachining process, the silicon-on-glass (SOG) process combines anodic bonding and deep reactive-ion etching (DRIE) techniques. The fabrication procedure usually needs only two or three lithography masks. Thus, it is relatively simple and low-cost fabrication process. In addition, all the technologies used in the process have been mature and well-established. In addition, the process has no risk of stiction, since no wet process is used for releasing the movable structures. Therefore, this process can fabricate large proof mass and capacitors and it can precisely control the geometry of the microstructures. The mechanical structures are made of single-crystal silicon, which has exceptional mechanical performances, such as low residual stress, good fatigue resistance and high yield strength.

Recently, the SOG-MEMS process has been utilized for gyroscopes and accelerometers [16, 17]. This is because it can minimize parasitic capacitance to improve sensitivity [18, 19]. The use of glass wafer as a substrate instead of silicon can make the parasitic capacitance minimal for not forming any silicon-oxide-silicon structures. Thus, this unique fabrication process could be a potential candidate of designing MEMS VR charge sensors. Moreover, the optimization of structure parameters of SOG-MEMS charge sensors is vital to achieve a high, linear, and wide dynamic range of charge sensitivity.

Pyrex 7740 glass wafer as a substrate and heavily doped silicon as a device layer are used to fabricate the electrometer in the SOG-MEMS process. The full fabrication process flow is illustrated in Fig. 3. Firstly, a 300 nm thick Ti/Pt/Au metallization layer (40/30/230 nm) is sputtered on the Pyrex glass wafer and then patterned using lift-off process to form electrical pads, as shown in Fig. 3a. To ensure the quality of anodic bonding, a self-alignment wet etch (such as buffered oxide etch) is used to recess the substrate by 250 nm before the metal sputtering. Thus, most of the metal is embedded in the shallow trench on the glass substrate, leaving 50 nm of the metal above the glass substrate. A 6-inch single-crystal silicon (SCS) wafer (N-type, <100>, 0.002 Ω·cm) is utilized for the fabrication of sensor structure. A 4 μm shallow silicon

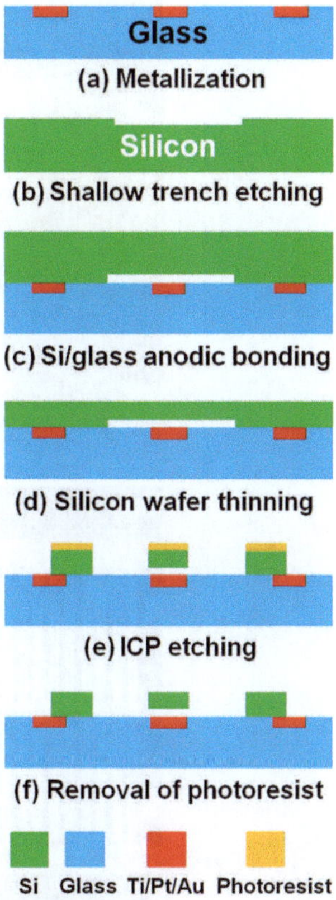

Fig. 3 Silicon-on-glass (SOG) MEMS fabrication process [19]. Copyright 2018. Reproduced with permission from IEEE Electron Device Letters

trench is etched using inductively coupled plasma (ICP) etching in SPTS OMEGA RAPIER, as shown in Fig. 3b. Si and glass wafers are aligned for anodic bonding in EVG510 wafer bonder (1000 V, 400 °C, 1 atmosphere pressure, 60 min), as shown in Fig. 3c. Then, Si wafer is thinned to approximate 32 μm by KOH etching, as shown in Fig. 3d. Following, the sensor structure is defined by photolithography and etched using high aspect ratio ICP etching, as shown in Fig. 3e. Finally, the photoresist is removed from the device using oxygen plasma etching, as shown in Fig. 3f. The photograph of the final fabricated MEMS vibrating-reed electrometer using Silicon-on-Glass technology is shown in Fig. 4.

4 Structural Design

The schematic of MEMS variable sense capacitor structure, as shown in Fig. 5, is constructed from an array of fixed electrodes, inter-digitally coupled to movable electrodes extending from the resonator shuttle. This array of equal-gap parallel-plate capacitors forms a balanced capacitive structure required for producing an

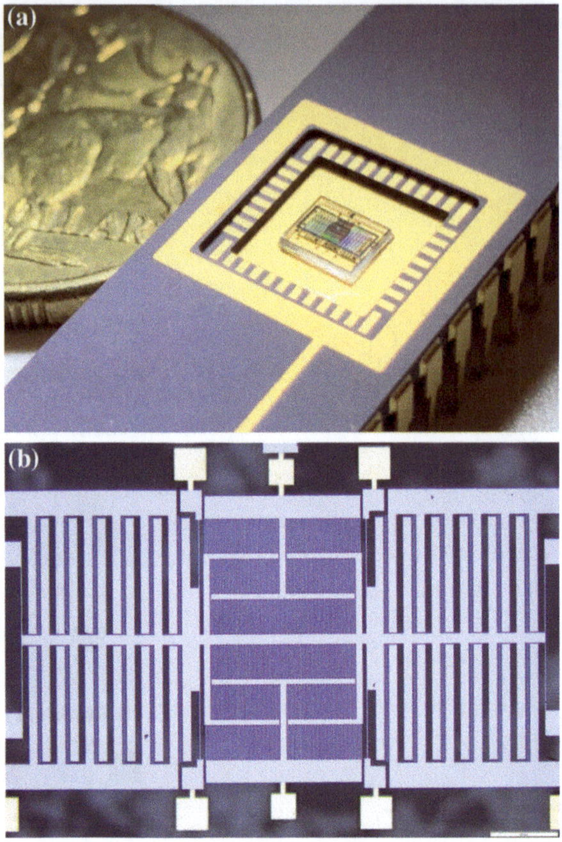

Fig. 4 Photographs of the silicon-on-glass (SOG) MEMS vibrating-reed electrometer **a** with package; **b** naked die

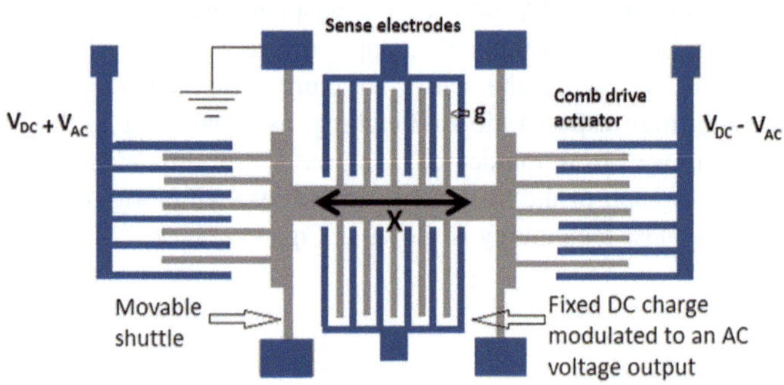

Fig. 5 Schematic of the MEMS variable capacitor as a resonator, actuated by superimposed DC and AC voltages

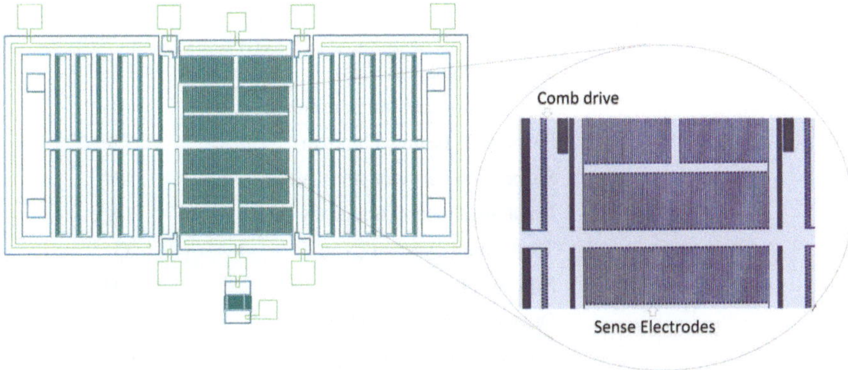

Fig. 6 The sensor layout design and a photograph of the fabricated MEMS electrometer (focused on the sensing electrode)

output with only even harmonics when driven harmonically at a fixed frequency. The modulated variable capacitor of the micromachined vibrating-reed electrometer is constructed from a resonator that is actuated laterally by comb drives on both sides, as shown in Fig. 5.

The details of the electrometer structural design are shown in Fig. 6. Here, the mask design layout of the MEMS device was prepared using Blueprint in Intellisuite software. The moving shuttle is supported by eight flexures (clamped-clamped guided silicon beams), which ensures the shuttle vibrate only in left-right direction as shown in Figs. 5 and 6. The complete device operates as a micromachined resonator, whereby an air gap capacitance is varied by changing the spacing between the fixed and movable electrodes.

A simplified way of modeling the proposed micromachined moving system is to form a lumped element model that contains a minimum number of physical parameters. A resonator (as shown in Fig. 5) vibrates as a response to an external force can be described using effective spring constant, effective mass and damping constant. Spring constant depends on the chosen material, the design structure, and in resonant cases also on the particular vibrational mode. Since a total of eight guided beams are used, thus the spring constant of the proposed sensor system can be expressed as

$$k = \frac{8EhW^3}{L^3} \tag{3}$$

where E is Young's modulus of the silicon material, L is the length of the beam, W is the width of the beam, and h is the thickness of the beam. The device structure parameters are summarized in Table 1. In this design, E is 165 GPa, W is 3.8 μm, h is 32 μm, and L is 404 μm. Thus, the total spring constant k can be calculated as 33 N/m.

Table 1 SOG-MEMS device design parameters and values

Parameter	Value
Structure thickness (μm)	32
Mass of movable structure (nKg)	105
Suspension beam width (μm)	3.8
Suspension beam length (μm)	404
Undamped natural frequency (kHz)	2.85
Total spring constant (N/m)	33
Number of sense combs per side	144
Drive capacitor gap, d (μm)	3
Number of drive combs per side	684
Drive comb length (μm)	20.5
Sense capacitance (pF)	2.49
Sense beam width (μm)	3.7
Sense capacitor gap, g (μm)	5.6

The mechanical damping factor or coefficient in parallel plate is given by

$$b = \frac{N_s \eta_{eff} L h^3}{g^3} \tag{4}$$

where N_S is number of sense electrodes, η_{eff} is air viscosity, sense electrode gap g is 5.6 μm. Accordingly, the calculated damping coefficient is 240 Ns/m.

The quality factor (Q factor) is the reciprocal of damping factor and is expressed as

$$Q = \frac{\sqrt{km}}{b} \tag{5}$$

In this design, the goal is to design an optimized variable sense capacitor, whose nominal value should be higher than parasitic capacitance of the MEMS device. On the other hand, a higher sense capacitance may increase the effective mass of the resonator due to more electrode plates required, thereby lowering natural frequency considerably. Thus, a higher spring constant is desired for the larger mass, ensuring that the natural frequency of the moving structure is above the corner frequency of Flicker noise in the sensing readout circuit.

With the application of Newton's second law, an electrostatic resonator can be regarded as a forced linear mass-spring-damper system with single-degree-of-freedom. As illustrated in Fig. 7, the system can be expressed as

$$m\frac{d^2x}{dt^2} + b\frac{dx}{dt} + kx = F_0 \, sin \, \omega \, t \tag{6}$$

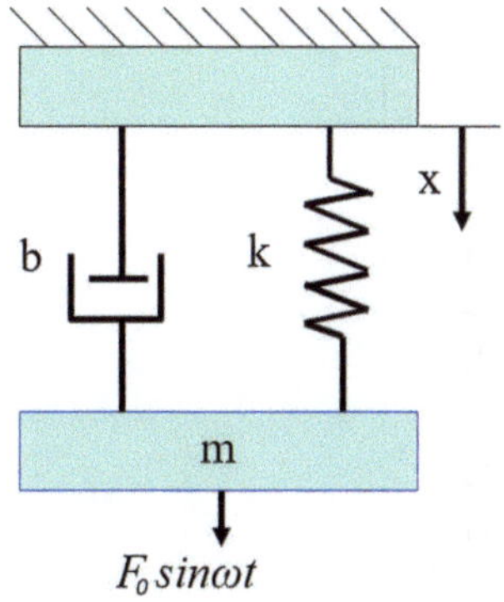

Fig. 7 Simplified model of a forced linear mass-spring-damper system for the single-degree-of-freedom resonator

Here, F_0 and ω are the maximum magnitude and frequency of the applied external force, respectively.

With Laplace transform notation, assuming zero initial conditions, the above equation can be converted to a second-order transfer function:

$$\frac{X(s)}{F(s)} = \frac{1}{ms^2 + bs + k} \tag{7}$$

The magnitude of the displacement can be represented as

$$|X(s)| = \frac{F_0/(m\omega^2)}{\sqrt{\frac{\omega_n^2}{\omega^2} - 1 + (2\zeta\omega_n)^2/\omega^2}} \tag{8}$$

where ω_n ($= \sqrt{k/m}$) is the natural frequency of the moving shuttle, ζ is the damping ratio, ω is the angular driving frequency. Here, the natural frequency is calculated as 2.85 kHz since effective mass was estimated as 105 nKg. From Eq. (5), the quality factor is estimated as 8. Damping ratio, which is inversely proportional to quality factor, can be defined as

$$\zeta = \frac{1}{2Q} \tag{9}$$

Using Eq. (9), the damping ratio can be calculated as 0.0625. Combining Eqs. (7), (8) and (9), the shuttle displacement frequency response can be derived as shown in Fig. 8. Here, $V_{AC} = 3$ V and $V_{DC} = 30$ V are applied to the push-pull comb drives, consequently, a maximum displacement of 2.52 μm is obtained at a resonant frequency of 2.85 kHz.

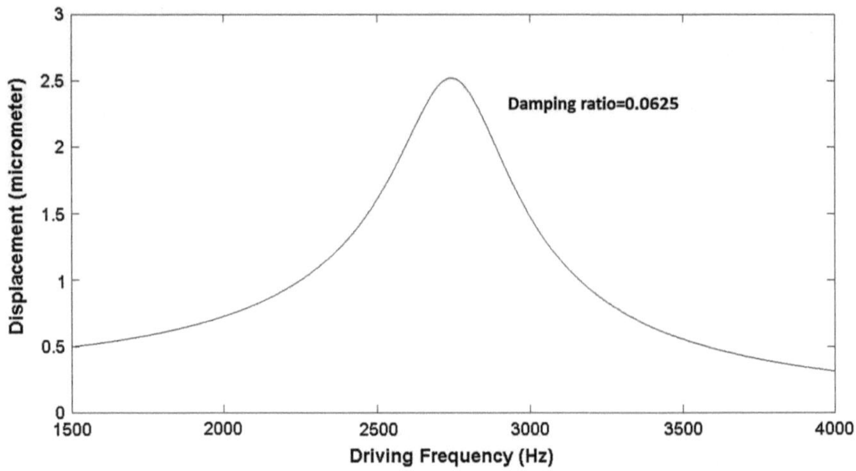

Fig. 8 Frequency response of the shuttle displacement under the applied electrostatic force

5 Circuit Topology Design

The key targets for opamp-based preamps design are: to minimize their noise, to reduce drifting, and to achieve sufficient gain. In addition to the intrinsic noise sources of opamps, charge measurements made with an electrometer are subject to a few DC error sources from the preamp, including an input leakage current I_b and input offset voltage V_{os}. Unavoidable leakage currents are ranging from a few fA to hundreds of nA from input transistors of the opamp, such as a junction field effect transistor (JFET), metal-oxide semiconductor field effect transistor (MOSFET), and bipolar junction transistor (BJT). As shown in Fig. 9, if the preamp is directly coupled with the sensor's output, the output of the vibrating capacitor integrates the leakage currents from that preamp, which is seen as a long-term drift in the electrometry. For example, at a fixed temperature, if the input leakage current is 10 pA, it causes a change in the charge detection of 10 pC per second. At extremely low charge levels of ~fC, even the small valued current might cause a significant error factor measuring the charge. If the input leakage current is known, it is possible to compensate for this error by subtracting the DC charge drift, due to the leakage current from the actual charge detection reading. Nonetheless, precisely determining the leakage current of the entire preamp system might not be an easy task during the measurement. To overcome this issue, a DC blocking or isolation capacitor C_C has been introduced between a sensing device and a preamp, so that the electrons are barred from the opamp's input to the sense electrodes during the charge measurement, as shown in Fig. 9.

In this design, a non-inverting, voltage-feedback opamp with single-pole system has been employed as a voltage-mode preamp, along with a SOG-MEMS charge sensor C_D, as shown in Fig. 9. Here, a gain resistor R_1 and a feedback resistor R_2 are

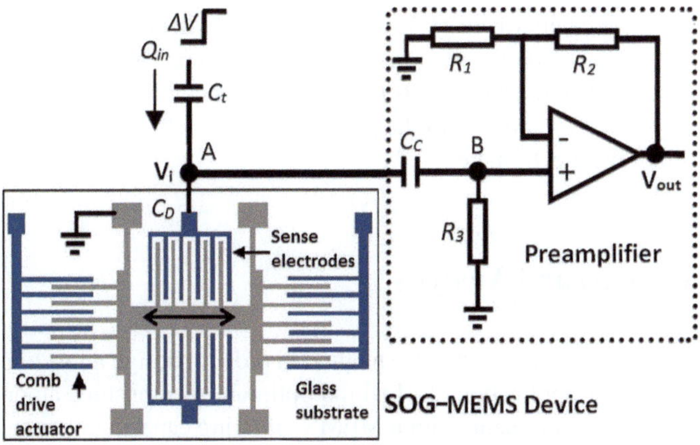

Fig. 9 Schematic of non-inverting voltage-mode preamp for SOG-MEMS vibrating-reed electrometer [10]. Copyright 2019. Reproduced with permission from IEEE Transactions on Instrumentation and Measurement

chosen to provide a voltage gain of the preamp. The input voltage step (ΔV) can be applied across the off-chip test capacitor C_t to set the injected charge ($\Delta Q_{in} = \Delta V \cdot C_t$) at the node A. For non-inverting voltage-mode preamps, there are two possibilities to reset the DC leakage currents into the input (node B in Fig. 9) of the opamp during the measurement process. These reset options could be implemented by offering a DC (or shunt) path either by switching (non-continuous) mode or by switch-less (continuous) mode. An electronic switch can be used for the switching mode DC path. When the transistor-based switch is closed, it couples the non-inverting input of the opamp to the ground. Therefore, it initializes the charge that may have been stored on the plates of the capacitor at the node A. However, this switching shunt circuit requires an external complex control circuitry. In addition, the switch might result in both parasitic leakage currents and a charge injection, which could be greater than the leakage currents of opamps. In Fig. 9, a passive shunt resistor of R_3 has been introduced to provide a direct current path. In the absence of R_3 at the input terminal of the opamp, the charge due to the intrinsic leakage current from the opamp starts to build on the input capacitor of the opamp, the stray capacitance (due to the wire and the circuit board), and the external non-polarized DC isolation capacitor C_C. This accumulation of unsolicited charge floats the DC potential at the input of the opamp, and eventually may drive it to voltage saturation and make the preamp circuit nonoperational. The AC coupling network of R_3C_C is placed between a MEMS sensing device and an opamp as shown in Fig. 9. The fixed non-polarized isolation capacitor C_C electrically isolates charge node A from input node B of the opamp blocking the leakage current to flow towards node A.

For an opamp as an amplifying device, an initial voltage offset at its input may induce a corresponding offset charge across the vibrating capacitor without inputting a continuous amount of charge to the node A (as shown in Fig. 9). Hence, a very low voltage offset at the input of opamp is essential to reduce such an initial offset at the output for the betterment of measurement error. However, in the design, both the leakage compensation and the offset correction circuits are circumvented. This is because the electronic components of these circuits contribute more RTI noise at the input of preamp in addition to the preamp circuit itself, and degrades the noise performance of the electrometer.

6 Noise Model and Analysis

A circuit noise analysis is an effective method in order to design a low noise preamp for a certain sensor application. To find the optimum relationships among the elements of the proposed preamp with a MEMS vibrating capacitor, a comprehensive circuit noise model and analysis must be carried out. In practice, for a well-designed MEMS VR electrometry system, the noise floor is entirely dominated by the noise characteristics of the preamp. Previously, commercial opamp-based preamps used for SOI-MEMS VR charge sensor described in [14, 15] did not adequately address all the design criteria to minimize the RTI noise voltage of e_{inn}. Zhu et al. [14] disregarded the equivalent current noise of opamp in their inverting voltage-mode preamp circuit. In this circuit, the equivalent input noise current flows through the sensor's capacitance and the AC coupling components resulting in an associated noise voltage due to that current noise. Thus, neglecting this noise contributor will cause an error during charge resolution calculation. In their noise model, Jaramillo et al. [15] neither considered the vibrating capacitor with all the parasitic capacitances of both the sensing device and the opamp, nor included the resistance of switching transistor (at off-mode), and subsequently overlooked the noise contribution due to the equivalent input noise current. Hence, it is necessary to identify and quantify all the component's noise sources of the preamp and account for them while calculating the overall noise of the front-end circuit.

In all band of frequencies, the equivalent input noise voltage e_n and current i_n of the opamp (originating from their internal components) are the combinations of thermal noise (Johnson noise), shot noise, and Flicker noise [11, 20]. Unlike Flicker noise, both Johnson noise and shot noise are independent of frequency. In amplifying devices, Johnson noise occurs from thermal fluctuations in the voltage across a resistor, whereas shot noise stems from the time-dependent fluctuations in electrical current due to quantized charges crossing tunnel junctions, p-n junctions and Schottky barrier diodes at random times [11]. It is known that the shot noise of opamps in a 1 Hz bandwidth is proportional to the square root of the leakage current, i.e.

$$i_n = \sqrt{2qI_b} \tag{10}$$

where q is the electron charge of 1.6×10^{-19} C [21].

Even though the leakage current in this design has been barred by the isolation capacitor C_C, it is important to choose an opamp which has a low valued leakage

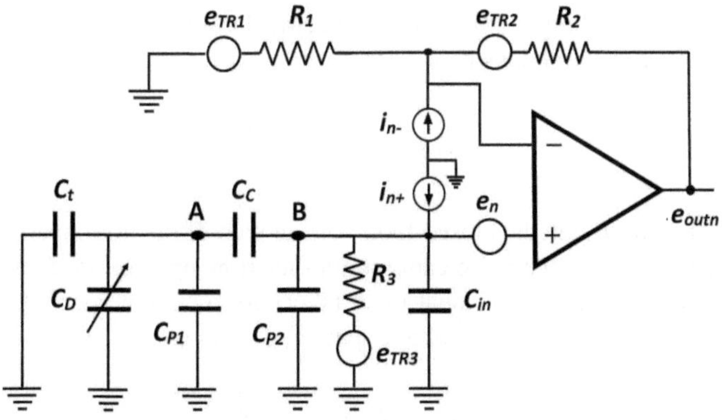

Fig. 10 Noise equivalent circuit of Fig. 9 including all noise contributions and parasitic capacitances [10]. Copyright 2019. Reproduced with permission from IEEE Transactions on Instrumentation and Measurement

current. The reason is while the leakage current of an opamp increases, the shot noise increases as well. The main noise sources contributed for a monolithic opamp-based voltage-mode premp are opamp input noise voltage e_n, inverting and non-inverting input noise currents (i_{n+}, i_{n-}) and thermal noises of external resistors (e_{TR1}, e_{TR2}, e_{TR3}) as illustrated in Fig. 10. It is worth mentioning that ideal capacitors do not yield thermal noise [20]. Although the noise contribution of a real capacitor is usually negligible, the input noise current i_n flowing through the capacitor develops the noise voltage. For a voltage-feedback opamp, the noise currents in the inverting and noninverting inputs are uncorrelated, i.e. they are two physically independent noise current sources but roughly equal in magnitude. Therefore, matched input transistors in voltage feedback opamps usually yield the same magnitude for both the noise current sources.

The total referred-to-output (RTO) noise voltage at the frequency of interest of the preamp is computed as the square root of the squared noise contributing terms (also known as the noise power density) to the output node. Because all the noise sources are treated as mutually exclusive, the computation is to add all the contributing noise power density at the output node by superposition. The square root is then taken to get back to that of the RTO noise voltage at the frequency of interest. Figure 10 represents a noise equivalent circuit of Fig. 9, including a noiseless opamp and associated resistors, along with MEMS capacitive device C_D and its parasitic capacitance of C_{P1}, test capacitor C_t, input capacitance of the opamp C_{in} and circuit board layout stray capacitance C_{P2}.

As showed in Fig. 10, there are totally six noise sources which can be identified and calculated individually at the output node e_{outn} as follows

1. Opamp equivalent input noise voltage spectral density e_n and the corresponding noise voltage spectral density is

$$e_{n01} = e_n\left(1 + \frac{R_2}{R_1}\right) \tag{11}$$

where $(1 + R_2/R_1)$ is the closed-loop voltage gain of the non-inverting opamp. The signal gain and the noise gain for non-inverting opamps are equal.

2. Opamp non-inverting equivalent input (shot) noise current spectral density i_{n+} and the corresponding noise voltage spectral density is

$$e_{n02} = \left(1 + \frac{R_2}{R_1}\right)\sqrt{\frac{R_3^2 i_{n+}^2}{1 + R_3^2 C_{total}^2 \omega^2}} \tag{12}$$

where C_{total} is equal to $(C_t + C_D + C_{P1})C_C/(C_t + C_D + C_{P1} + C_C) + C_{in} + C_{P2}$.

3. Opamp inverting equivalent input (shot) noise current spectral density i_{n-} and the corresponding noise voltage spectral density is

$$e_{n03} = i_{n-} R_2 \tag{13}$$

4. Shunt resistor R_3 equivalent input thermal noise voltage spectral density e_{TR3} and the corresponding noise voltage spectral density is

$$e_{n04} = \left(1 + \frac{R_2}{R_1}\right)\sqrt{\frac{4kTR_3}{1 + R_3^2 C_{total}^2 \omega^2}} \tag{14}$$

where k is Boltzmann's constant ($1.38064852 \times 10^{-23}$ J/K), and T is temperature in Kelvin. It is noted that R_3C_C network acts as a high-pass filter (HPF) with respect to the modulated charge signal to be measured, while R_3C_{total} acts as a low-pass filter (LPF) with respect to the thermal noise generated by the shunt resistance R_3.

5. Gain resistor R_1 thermal noise voltage spectral density e_{TR1} and the corresponding noise voltage spectral density is

$$e_{n05} = \frac{R_2}{R_1}\sqrt{4kTR_1} \tag{15}$$

6. Feedback resistor R_2 thermal noise voltage spectral density e_{TR2} and the corresponding noise voltage spectral density is

$$e_{n06} = \sqrt{4kTR_2} \tag{16}$$

Here, it is considered as $|i_{n+}| = |i_{n-}|$. By analyzing Fig. 10, the RTO noise voltage can be expressed as follows

$$e_{outn} = \sqrt{e_{n01}^2 + e_{n02}^2 + e_{n03}^2 + e_{n04}^2 + e_{n05}^2 + e_{n06}^2} \tag{17}$$

The voltage transfer function from node A to node B can be represented as

$$\frac{V_B}{V_A} = \sqrt{\frac{\omega^2 R_3^2 C_C^2}{1 + \omega^2 R_3^2 (C_C + C_{in} + C_{P2})^2}} \tag{18}$$

Thus, the system voltage gain from node A to node e_{outn} can be defined as

$$A_v = \left(1 + \frac{R_2}{R_1}\right)\sqrt{\frac{\omega^2 R_3^2 C_C^2}{1 + \omega^2 R_3^2 (C_C + C_{in} + C_{P2})^2}} \tag{19}$$

The total referred-to-input (RTI) noise voltage (e_{inn}) per square root Hz of the preamp circuit is calculated as follows

$$e_{inn} = \frac{e_{outn}}{A_v} \tag{20}$$

Finally, assuming $\omega^2 R_3^2 (C_C + C_{in} + C_{P2})^2$ is much larger than 1, e_{inn} at node A can be expressed by combining equations of (17), (19) and (20),

$$e_{inn} = \left(1 + \frac{C_{in} + C_{P2}}{C_C}\right) \left[e_n^2 + \frac{i_n^2}{C_{total}^2 \omega^2} + \frac{4kT}{R_3 C_{total}^2 \omega^2} + i_n^2 (R_1||R_2)^2 + 4kT(R_1||R_2)\right]^{\frac{1}{2}} \tag{21}$$

As can be seen from Eq. (21), all five noise terms depend on a scale factor of $\left(1 + \frac{C_{in}+C_{P2}}{C_C}\right)$. Hence, reducing the value of input capacitance C_{in} and parasitic capacitance C_{P2} of the opamp and increasing the value of isolation capacitance C_C can reduce e_{inn} considerably. According to the first, second and fourth terms in the square bracket in Eq. (21), e_{inn} can be reduced for a low value of e_n and i_n of an opamp by selecting low input noise opamps. To reduce the effect of the second and third noise terms of Eq. (21), the operating frequency of the sensor needs to be higher. In this regard, the resonance frequency f of the sensor's movable electrode must be in a higher range. For femto-ampere of i_n, the fourth term of Eq. (21) can be neglected due to its insignificant magnitude of femto-V/rtHz.

At first glance of the preamp circuit as shown in Fig. 10, one might think that R_3 needs to be lowered because higher resistance contributes higher thermal noise. However, to minimize the noise voltage of the third noise term of Eq. (21), the value of R_3 should be maximized in such a way that the output of the preamp is not being saturated for a fixed value of C_C. For example, 10 pA of leakage current with 100 GΩ of R_3 yields 1 V developed across the input of opamp which can potentially cause a complete failure of detecting charge signals from the sensor. Without any input signal, this unwanted DC output error may exhibit a finite output voltage (V_{error}). To ensure this output signal is not saturated, closed-loop gain of the opamp, leakage current, and offset voltage must be low, and the following design condition needs to be considered

$$R_3 \leq \frac{V_{error} - V_{os}\left(1 + \frac{\mathrm{R}_2}{\mathrm{R}_1}\right)}{I_b\left(1 + \frac{\mathrm{R}_2}{\mathrm{R}_1}\right)} \tag{22}$$

By selecting very low valued resistances for both gain and feedback resistors, the fourth and fifth noise terms can be made inconsequential for designing a low noise preamp. The examination of Eq. (21) shows that R_3 and C_C are the two main parameters that can be manipulated to minimize noise for a given sensor and amplifying device.

Using the fixed values in Table 2, the total equivalent RTI noise voltage (e_{inn}) is calculated using Eq. (21) in MATLAB. As shown in Fig. 11, e_{inn} is a function of R_3 (from 100 kΩ to 10 GΩ) and C_C (from 0.1 to 3 pF). By analyzing the graph, a significantly high noise voltage is noticed for lower values of C_C and R_3. On the other hand, for $C_C > 1.5$ pF and $R_3 > 10$ MΩ, the noise voltage reduces remarkably. The above analysis indicates that considering Eq. (21), higher pF range of C_C and higher GΩ range of R_3 are desirable attributes for designing a low noise preamp. According to Eq. (21), further noise optimization of preamp can be attained by selecting an opamp having lower input noise of e_n and i_n.

By analyzing Figs. 9 and 10 shown previously, the sensitivity for VR electrometer is given as

Table 2 Noise calculation parameters and values

Parameter	Value
i_n	0.6×10^{-15} A/rtHz
e_n	7.5×10^{-9} V/rtHz
f	5.2 kHz
k	1.38×10^{-23} J/K
T	298 K
C_{in}	2.5 pF
C_t	2 pF
C_D	4 pF
C_{P1}	1 pF
C_{P2}	1 pF
R_1	49.9 Ω
R_2	680 Ω

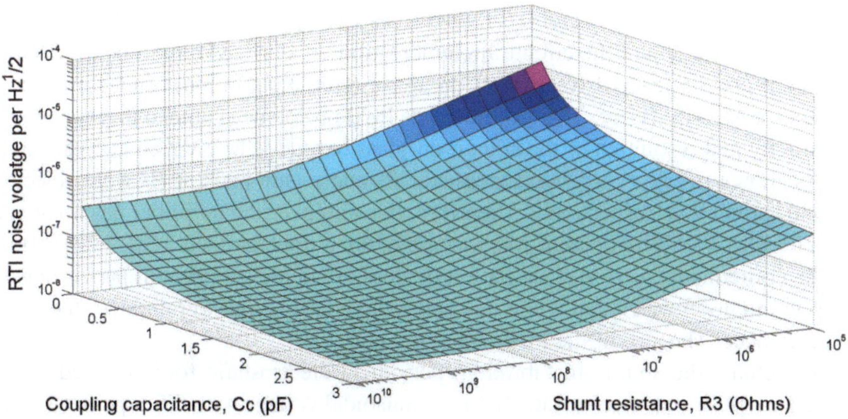

Fig. 11 Calculated RTI noise voltage at 5.2 kHz using Eq. (21) as a function of C_C and R_3 [10]

$$S_e = \frac{C_D}{2\sqrt{2}\left[C_D + C_t + C_{P1} + \frac{C_C(C_{in}+C_{P2})}{C_C+C_{in}+C_{P2}}\right]^2}\left(\frac{x}{g}\right)^2 \tag{23}$$

where x/g is the vibrating plates traveling range ratio of the MEMS device. Finally, using Eqs. (1), (21) and (23), the charge resolution of the proposed electrometer is represented as

$$Q_n = \frac{e_{inn}}{8\sqrt{2}C_D}\left[C_D + C_t + C_{P1} + \frac{C_C(C_{in} + C_{P2})}{C_C + C_{in} + C_{P2}}\right]^2 \tag{24}$$

where the vibrating plates traveling range x is half of the initial gap g.

7 Experimental Setup and Results

The optimized structural design of the MEMS electrometer was manufactured in a Silicon-on-Glass (SOG) standard process, as discussed in Sect. 3. A 28-pin dual-in-line (DIP) open cavity ceramic package was used to house the sensor die, as illustrated in Fig. 4a. Gold ball wire bonding was utilized to electrically interconnect the sensor with interface circuity. Actuation and sensing circuits are custom-made on a PCB, including push-pull driver and preamplifier as discussed in Sect. 5. The experimental testing was carried out in standard electronic laboratory with general equipment including power generator, arbitrary function generator, oscilloscope, spectrum analyzer, etc.

Based on the noise analysis in Sect. 6, a low noise opamp of AD8067 was selected to form the preamplifier for the electrometer. According to the datasheet of AD8067,

e_n and i_n are 7.5 nV/rtHz and 0.6 fA/rtHz, respectively at 5.2 kHz [22]. Figure 12 shows a photograph of the circuit implementation on a printed circuit board (PCB). The PCB prototype has driver circuit, preamp circuit, input/output connectors, and a socket for MEMS device. The custom-made PCB prototype is able to investigate the sensitivity and resolution of the silicon-on-glass based MEMS VR charge sensor.

As shown in Fig. 9, a feedback resistor R_2 and a gain resistor R_1 were selected as 680 and 49.9 Ω, respectively. Therefore, this noninverting amplifier has a voltage gain of 14.6 V/V. The preamp was assembled on a double-sided PCB, where most of the components were surface mount devices. To achieve optimum performance, careful attention was given when designing the board layout to minimize the stray capacitances. The size of the PCB is 8×4.4 cm^2.

To actuate the shuttle to vibration, periodical electrostatic force is needed. To generate a push-pull electrostatic force, a sinusoidal voltage with 5.5 V peak-to-peak magnitude at 2.6 kHz and a DC polarization voltage of 26 V were superimposed. The actuation voltages were applied to drive the SOG-MEMS charge device at a room temperature of 298 K and ambient pressure so that the displacement amplitude (x) of the resonator reaches approximate half of the vibrating capacitor gap (g). The larger displacement ensures the larger charge sensitivity. However, one must make sure the displacement is less than the capacitor's initial gap, otherwise the resonance will stop. As shown in Fig. 13, the displacement can be visibly observed through microscope, evidenced by the blurry structures while vibrating.

To apply a known amount of charge, a test capacitance C_t (2 pF) was utilized to induce charge by applying a step voltage on it. Incremental levels of step voltages were applied to this capacitor using a voltage source. Steps of 0.1 V to C_t were calibrated to the amounts of charge on the fixed terminal of the sensing device, which was the input terminal of the electrometer, as shown in Fig. 9. A mixed signal oscilloscope (MSO-X 3104A, Agilent Technologies) was utilized in the fast Fourier transform (FFT) mode to measure the output signal from the preamp. The second

Fig. 12 A photograph of the preamp and the driver circuit on the PCB for SOG-MEMS vibrating-reed electrometer [10]. Copyright 2019. Reproduced with permission from IEEE Transactions on Instrumentation and Measurement

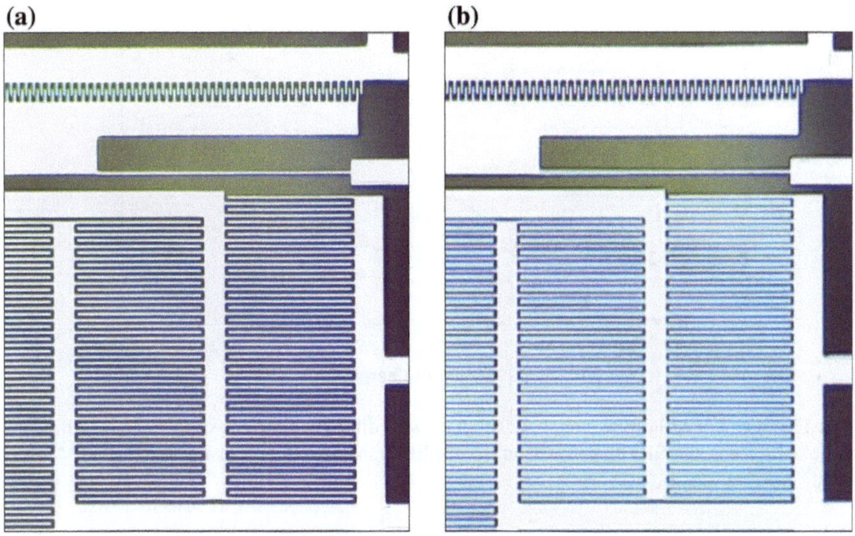

Fig. 13 Local structures of the sense and drive structures in the micro-machined electrometer. **a** The static device with no actuation voltage applied; **b** the vibrating device with visible vibration (blurry structures) at 2.8 kHz with displacement of approximate half of the capacitor gap

harmonic of the signal is the signal corresponding the input charge, as shown in Fig. 14. As the input charge increases, the second harmonic peak voltage increases.

It should be mentioned that this output voltage V_{out} has been amplified by the system voltage gain of A_v. Thus, to calculate the charge-to-voltage conversion gain,

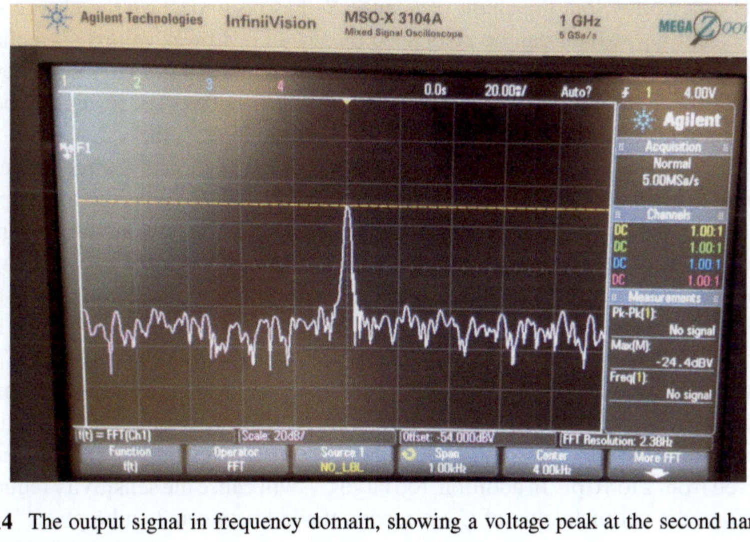

Fig. 14 The output signal in frequency domain, showing a voltage peak at the second harmonic frequency

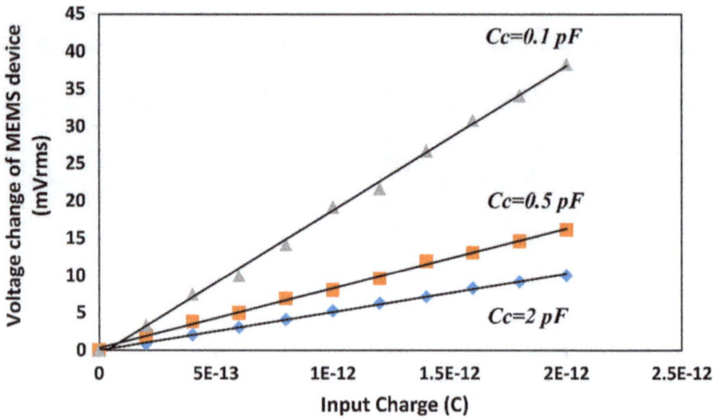

Fig. 15 SOG-MEMS electrometer sensitivities with different coupling capacitances [10]. Copyright 2019. Reproduced with permission from IEEE Transactions on Instrumentation and Measurement

this voltage gain needs to be removed during the calculation. Keeping the opamp and the resistors' parameter unchanged, Fig. 15 shows the linear relationships between the unamplified (rms) voltage and the input charge with respect to three different isolation capacitances of 0.1, 0.5 and 2 pF. The sensitivities of 5.06×10^9 and 8×10^9 V/C were achieved with the isolation capacitances of 2 and 0.5 pF, respectively. On the other hand, the highest sensitivity of 1.9×10^{10} V/C was achieved with 0.1 pF isolation capacitance (C_C). It was observed from the readings in Fig. 15 that the charge sensitivity increased by reducing the value of the isolation capacitor. The reason is that the larger isolation capacitance, the more charge accumulates on it instead of the sensing capacitor, thereby reducing the output voltage.

It is noteworthy that the shunt resistance R_3 has no impact on the charge sensitivity except C_C, which can be evidenced by Eq. (23). Lower valued C_C enhances the sensitivity based on Eq. (23) and verified experimentally in Fig. 15. However, it increases the noise contribution based on Eq. (21) and affects the resolution performance based on Eq. (24). To overcome these predicaments, the value of C_C must be increased in such a way that the sensitivity remains high, and the preamp circuit contributes low noise to achieve a high resolution. In order to determine the optimal value of C_C, the charge resolution Eq. (24) has been numerically simulated, and the relationship between charge resolution and blocking capacitance is shown in Fig. 16.

To reveal a bigger picture of this relationship, range of C_C is selected from 0.1 to 10 pF. Here, we can see that the resolution value is almost constant when C_C is larger than 5 pF. This can be explained by Eq. (24): because C_C is in parallel with C_{in} and C_{p2}, when C_C is larger than $(C_{in} + C_{p2})$, the impact of C_C becomes less dominant. Thus, for higher valued C_C, the resolution keeps almost no change. According to the simulation results shown in Fig. 16, no significant resolution changes can be observed from 2 to 10 pF. In addition, too large C_C will cause the sensitivity reduction. Therefore, the optimal value of C_C is recommended between 2 and 6 pF.

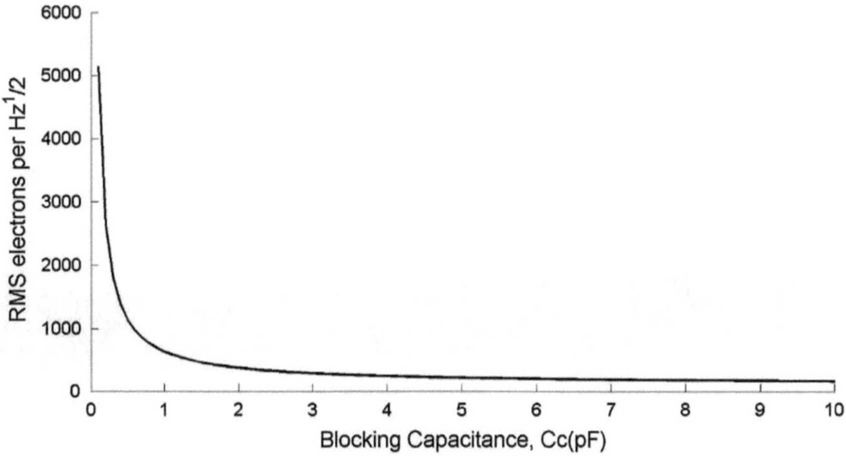

Fig. 16 Charge resolution versus blocking capacitance C_C

Apart from C_C, the test capacitor C_t is also a controllable parameter and has an impact on resolution based on Eq. (24). According to Eqs. (12) and (14), C_t contributes to the overall noise, and according to Eq. (24), C_t does contribute to the charge resolution. Therefore, C_t should be selected as small as possible to enhance the charge resolution performance. During measurements in this work, C_t was included to supply a known charge value for sensor's characterization purpose. However, in practical applications, such as detection of bio-analyte or aerosol particles, C_t does not exist and can be replaced by a conductor or fixed electrode. Hence, this ancillary capacitor's impact is inconsequential in overall charge resolution.

Experimental noise measurement was performed to validate the noise model of the proposed circuit with C_C of 2 pF and R_3 of 50 MΩ. Attained results were cross-checked with simulation results. A simulated RTI noise voltage of 317 nV/rtHz was achieved by using a SPICE compatible circuit simulator of TINASPICE. An experimental RTO noise voltage of 3.07 μV/rtHz at 5.2 kHz was measured using a spectrum analyzer (USB-SA44B, Signal Hound), as shown in Fig. 17. The output noise measurement was performed in the frequency range from 1 Hz to 10 kHz. The overall gain of the preamp circuit was measured as 10.6 V/V. Thus, total RTI noise voltage was determined as 290 nV/rtHz at 5.2 kHz. The noise measurement and noise simulation results of the preamp confirm the theoretical noise model of Eq. (21) and gives a direction for further optimization of the circuit. To determine the charge resolution for $C_C = 2$ pF, a tested sensitivity of 5.06×10^9 V/C was considered, as shown in Fig. 15. Therefore, a charge resolution of 57.31 aC/rtHz or 358.2 electrons/rtHz was achieved.

Both the R_3-C_C network and the opamp, along with the capacitive MEMS device, have a significant impact on determining the RTI noise voltage of the electrometer. As shown in the graph of Fig. 18, the overall charge resolution was evaluated as a function of C_C and R_3 using Eq. (24) and parameters in Table 2. A significant difference of

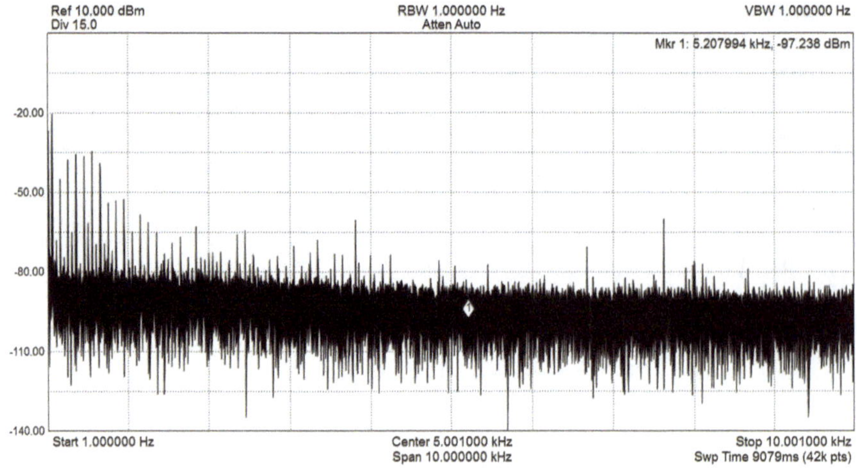

Fig. 17 The measured total RTO noise voltage of the preamp for $C_C = 2$ pF and $R_3 = 50$ MΩ [10]. Copyright 2019. Reproduced with permission from IEEE Transactions on Instrumentation and Measurement

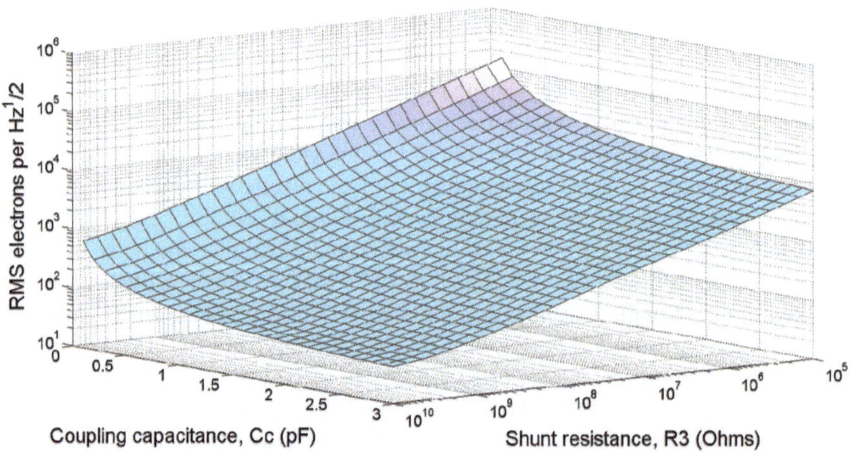

Fig. 18 Calculated charge resolution as a function of C_C and R_3 [10]. Copyright 2019. Reproduced with permission from IEEE Transactions on Instrumentation and Measurement

charge resolution can be observed for high values of C_C and R_3 in comparison with low values of C_C and R_3. For this designed preamp, Fig. 19 represents the noise contributions from all the noise generators using Eq. (21). It was quantified that 89% of the total noise contributor was the third term of Eq. (21) for R_3 of 50 MΩ, as shown in Fig. 19a. Thus, lower valued R_3, which is the dominant noise contributor of e_{inn}, reduces the overall charge resolution based on Eq. (24). If R_3 is increased to 20 GΩ, e_{inn} can be reduced significantly to 28 nV/rtHz due to the significant reduction of the

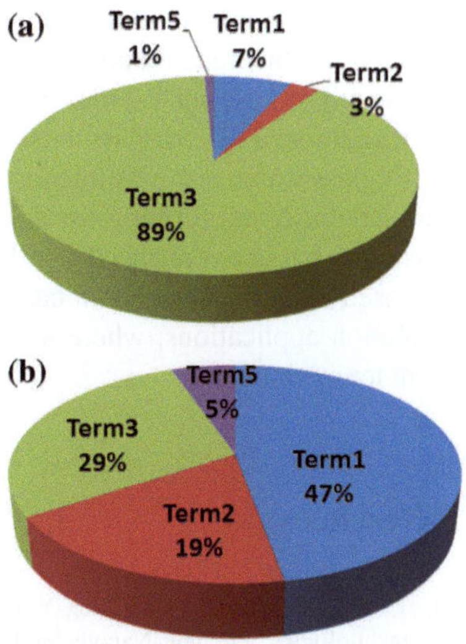

Fig. 19 Noise contributions of the preamp for **a** $R_3 = 50\ \text{M}\Omega$, **b** $R_3 = 20\ \text{G}\Omega$ [10]. Copyright 2019. Reproduced with permission from IEEE Transactions on Instrumentation and Measurement

thermal noise of R_3. In this case, it contributes 29% of the total noise, as shown in Fig. 19b. On the other hand, 47% of the total noise is due to the opamp noise voltage. Hence, for shunt resistor in GΩ range, further reduction of RTI noise voltage can be possible by selecting JFET-input opamps with very low e_n, i_n and C_{in}.

8 Conclusion

In this study, an optimal non-inverting voltage-mode preamplifier for charge sensors has been designed and implemented by analyzing various noise contributors of the shot noise and the Johnson-Nyquist noise at frequencies above the corner frequency of Flicker noise. With the development of the MEMS based vibrating-reed (VR) electrometer, both the sensitivity and a systematic analysis of the noise model for the first-stage of the readout circuit or the preamp have become crucial to quantify the charge resolution. By comparing analytical calculations and experimental measurements, the complete electrometry provides a vivid depiction of its performance rather than just stating the total result. Previously published descriptions of the preamp circuits have neglected a few vital parameters, which are important for evaluating the individual contributions to the charge signal response and the pramp noise.

The total referred-to-the-input (RTI) noise voltage, in this work, mainly consists of three components: opamp input-referred noise voltage e_n, non-inverting opamp input-referred noise current i_n, and noninverting shunt resistor R_3 thermal noise voltage e_{TR3}. Among them, the thermal noise voltage significantly contributes to high noise due to MΩ range of resistance, while the remaining noise components contribute comparably to lower noise. The noise floor detection method was studied for the SOG-MEMS vibrating-reed sensing device with an ultra-low noise AD8067 opamp and an RC network consisting of 50 MΩ and 2 pF. A good correlation was obtained between the theoretical and experimental noise outcomes within the measurement bandwidth of 1 Hz for 5.2 kHz at room temperature and ambient pressure.

In addition, a silicon/glass anodic bonding technology has been exploited, and the use of glass wafer as a substrate instead of silicon can make the parasitic capacitance minimal due to not forming any silicon-oxide-silicon structures. Hence, the sensor's sensitivity has been enhanced significantly in comparison with SOI counterparts. The proposed electrometer system can be used in portable instruments and high charge resolution applications, where less than 10 e/rtHz charge resolution is required at room temperature.

References

1. J. Jalil, Y. Zhu, C. Ekanayake, Y. Ruan, Sensing of single electrons using micro and nano technologies: a review. Nanotechnology **28**, 142002 (2017)
2. I. Ahmed, J.A. Haigh, S. Schaal, S. Barraud, Y. Zhu, C.M. Lee, M. Amado, J.W.A. Robinson, A. Rossi, J.J.L. Morton, M.F. Gonzalez-Zalba, Radio-frequency capacitive gate-based sensing. Phys. Rev. Appl. **10**, 014018 (2018)
3. R.J. Schoelkopf, P. Wahlgren, A.A. Kozhevnikov, P. Delsing, D.E. Prober, The radio-frequency single-electron transistor (RF-SET): a fast and ultrasensitive electrometer. Science **280**, 1238–1242 (1998)
4. A.N. Cleland, M.L. Roukes, A nanometer-scale mechanical electrometer. Nature **392**, 160–162 (1998)
5. J.E.Y. Lee, B. Bahreyni, A.A. Seshia, An axial strain modulated double-ended tuning fork electrometer. Sens. Actuator A Phys. **148**, 395–400 (2008)
6. D. Chen, J. Zhao, Y. Wang, Z. Xu, J. Xie, An electrostatic charge sensor based on micro resonator with sensing scheme of effective stiffness perturbation. J. Micromech. Microeng. **27**, 065002 (2017)
7. P.S. Riehl, K.L. Scott, R.S. Muller, R.T. Howe, J.A. Yasaitis, Electrostatic charge and field sensors based on micromechanical resonators. J. Microelectromech. Syst. **12**, 577–589 (2003)
8. Y. Zhu, J. Lee, A. Seshia, System-level simulation of a micromachined electrometer using a time-domain variable capacitor circuit model. J. Micromech. Microeng. **17**, 1059–1065 (2007)
9. Y. Zhu, J. Lee, A. Seshia, MEMS electrometer system simulation using a time-domain variable capacitor model, in *TRANSDUCERS 2007–2007 International Solid-State Sensors, Actuators and Microsystems Conference*, Lyon, 2007, pp. 1685–1688
10. J. Jalil, Y. Ruan, H.Z. Li, Y. Zhu, Comprehensive design considerations and noise modeling of preamplifier for MEMS electrometry. IEEE Trans. Instrum. Meas. (2019). https://doi.org/10.1109/tim.2019.2930440
11. S.M. Sze, K.K. Ng, *Physics of Semiconductor Devices* (Wiley, New Jersey, 2007)
12. J. Lee, Y. Zhu, A. Seshia, Room temperature electrometry with SUB10 electron charge resolution. J. Micromech. Microeng. **18**, 025033 (2008)
13. J.E. Lee, Y. Zhu, A.A. Seshia, A micromechanical electrometer approaching single-electron charge resolution at room temperature, in *2008 IEEE 21st International Conference on Micro Electro Mechanical Systems*, Wuhan, 2008, pp. 948–951
14. Y. Zhu, J.E.Y. Lee, A.A. Seshia, A resonant micromachined electrostatic sensor. IEEE Sens. J. **8**, 1499–1505 (2008)
15. G. Jaramillo, C. Buffa, M. Li, F.J. Brechtel, G. Langfelder, D.A. Horsley, MEMS electrometer with femtoampere resolution for aerosol particulate measurements. IEEE Sens. J. **13**, 2993–3000 (2013)

16. J. Wang, Z. Yang, G. Yan, A silicon-on-glass Z-axis accelerometer with vertical sensing comb capacitors, in *Proceedings of IEEE 7th International Conference on Nano/Micro Engineered and Molecular Systems (NEMS)*, Kyoto, May 2012, pp. 583–586
17. Z. Ma, Y. Wang, Q. Shen, H. Zhang, X. Guo, Key processes of silicon-on-glass MEMS fabrication technology for gyroscope application. Sensors **18**, 1240 (2018)
18. C. Lin, C. Hsu, H. Yang, W.C. Wang, W. Fang, Implementation of silicon-on-glass MEMS devices with embedded through-wafer silicon vias using the glass reflow process for wafer-level packaging and 3D chip integration. J. Micromech. Microeng. **18**, 025018 (2008)
19. J. Jalil, Y. Ruan, Y. Zhu, Room-temperature sensing of single electrons using vibrating-reed electrometer in silicon-on-glass technology. IEEE Electron Device Lett. **39**(12), 1928–1931 (2018)
20. C.D. Motchenbacher, J.A. Connely, *Low Noise Electronic System Design* (Wiley, New York, 1993)
21. E. Serrano-Finetti, R. Pallas-Areny, Noise reduction in AC-coupled amplifiers. IEEE Trans. Instrum. Meas. **63**(7), 1834–1841 (2014)
22. Analog Devices, Inc. Datasheet AD8067 [Online]. Available: http://www.analog.com/media/en/technical-documentation/datasheets/AD8067.pdf. Accessed 1 Apr 2017

Micromachined Resonator-Based Charge and Electric Field Sensors: A Review

Emad Esmaeili and Behraad Bahreyni

Abstract An electrometer is a sensor to measure electric charge. Electrometers are needed in various applications, ranging from the detection of ionization charges in nuclear physics, counting ions in mass spectroscopy, and space exploration, among others. Most electrometers measure the charge indirectly. For instance, solid-state electrometers measure the electric potential that is generated by an induced charge across the electrodes of a known capacitance. Devices such as gold-leaf electrometer, on the other hand, measure the columbic force between charges. Yet some other electrometers utilize a continuously varying capacitor to convert input charge to an AC current which is often easier to measure. Solid-state and vacuum-tube based electrometers have been developed as miniaturized, low-cost alternatives to traditional systems. These devices, however, suffer from drift, low-frequency noise, and leakage. Micromachined electrometers have been developed to address such short comings. Resonant sensing is often employed due to the need for resolving rather small forces from input charges. In this chapter, we will look at two main approaches for the design of micromachined electrometers, where the input charge affects the response of either a single micro-resonator or the combined response of coupled micro-resonators. We also discuss some micromachined devices for the measurement of electric field, as such devices in many applications can be utilized as electrometers.

1 Introduction and History

Electrometer is an electrical device to measure the small electric charge, current or electric potential difference. Electrometers have applications in space exploration includes the detection of electrostatic charges which are the source of dust motion in the dry Martian atmosphere [1]. In both tunneling and scanning electron microscopy systems, it is necessary to have a proper current for producing the picture of the samples [2]. In electron microscopy, electrometers are used to quantify the electron emission current by measuring the number of electrons that reach or cross the sample.

E. Esmaeili · B. Bahreyni (✉)
School of Mechatronic Systems Engineering, Simon Fraser University, Surrey, BC, Canada
e-mail: behraad@ieee.org

Y. Zhu (ed.), *Micro and Nano Machined Electrometers*,
https://doi.org/10.1007/978-981-13-3247-0_6

As the electric charge is a vital characteristic of biological molecules, mass spectrometry for chemical and biological characterization is one of the most important applications of electrometers [3, 4]. In nuclear research and ionization radiation measurement, an electrometer can be used to measure the produced electrical charges and act as a current to frequency converter [5]. Also, electrometer can be utilized in environment air quality monitoring where ultrafine particles can be detected and classified by exploiting the dependence of their electrical charge and their size [6, 7].

About 200 years ago, Charles Coulomb reported that electrostatic charges exert a force on each other. He invented the torsion balance which was the very first reported sensor to quantify charge. Millikan used an oil drop experiment to quantify the charge of a single electron. He charged an oil drop and used an electric field to prevent it from falling. By determining the electric field, he measured the charge. He repeated this experiment with oil droplets with different charges and was able to eventually measure a single electron charge. In 1932, Ross Gunn invented the first variable capacitor electrometer. His vibrating reed electrometer was based on the variable capacitance, which was created by semi-cylindrical electrodes, two stationary electrodes and two on a rotating shaft. Using this system, he was able to quantify a minimum charge of 4 fC. In recent years, exploiting CMOS technology and cryogenically cooled single-electron transistors, the charge resolution is reached to $10^5\ e/\sqrt{Hz}$ [8]. Although other solid-state devices included MOSFETs are able to detect single electrons at a higher temperature of 8 K [9], these methods are not still applicable for room temperature measurement. Commercially available electrometers have the capability to sense the equivalent charge of 50,000 electrons. For example, the benchtop Keithley 6517A is a solid-state electrometer with a minimum resolution of 10 fC ($\approx$ 63,000 electrons) [10]. Although these types of equipment are accessible, the resolution is not enough for scientific application. Preamp noise is the main reason that commercial devices have such moderate sensitivity. To meet the resolution sufficient for the scientific application, it is necessary to avoid the preamplifier noise which includes three common electrical noises: Johnson noise, shot noise and flicker noise in different bandwidths of operation [11, 12].

On the other hand, electric field sensors are the most similar devices to the electrometers. Electric field sensors are field measuring devices that often work on the electrostatic induction principle. They can be used to quantify both DC and AC eclectic fields. They are mainly used to detect and measure the electrostatic field intensity and electrostatic charges. Many different application areas such as atmospheric science, power systems, and high-tech equipment requires to measure and monitor the electric fields.

In industrial applications, it is necessary to measure and control the electric field intensity in order to prevent hazardous situations [13]. Electrostatic charges can be of importance in industries such as paper, plastic, and fuel storing and transporting containers.

In power systems, monitoring of the electric fields around the installations is required to improve the insulation systems in different weather and to ensure personal and utilities safety. Moreover, electric field sensors are used to detect the faulty insulators, measuring field near to high voltage lines and periodic verifications of insulator strings in power transmission [14, 15]. Using electric field sensors, one can measure voltage in three-phase power systems without using bulky and expensive insulating structures [16].

Another application for electric field sensors is the measurement of atmospheric electric fields in order to avoid lightning strikes to space vehicles. Moreover, different industrial applications can use these sensors in electrostatic discharge sensitive electronics [5–19]. These sensors also are used for ice accretion detection on high-voltage and electric transmission lines and transformers [20].

The increasing demand for more sensitive and accurate charge and electric field sensors for different application motivate the researchers to utilize microelectromechanical systems (MEMS) to fabricate smaller and more accurate devices. The ability of the MEMS technology to reduce the size of the device not only can increase the throughput but also it will reduce the cost per device. Other benefits include compatibility to electronic circuits, low power consumption, etc. which caused growing interest in using MEMS technology in the area of the sensor.

The first MEMS electrometer was reported by Riehl et al. [21]. They employed a vibrating MEMS capacitance electrometer using a resonating structure created by an electrostatic comb drive that is differentially actuated on both sides by and sense parallel-plate capacitors.

2 Micromachined Resonators

2.1 Electrostatic Transduction

Most of the MEMS electrometer and electric field sensors operate based on the dynamic behavior of the device which is the result of the device's proper excitation. Each resonator has its structure and geometry dependent resonance frequency and by exciting it in that frequency, the output would be an amplified response. Moreover, the damping which is basically the energy dissipation in the system would be a significant issue when the dynamic behavior of the system is a matter of importance. In resonator devices, the energy needed for vibration can be coming from different sources and mechanisms including electrostatic actuation, thermal actuation, piezoelectric actuation, etc.

Electrostatic forces are generated between structures with different voltage. They can be charged independently or proportionally. An electrostatic attraction force can be produced between the charged electrodes of a capacitor by applying a voltage across them. The magnitude of the force is related to the voltage difference between the electrodes, their geometry and the gap between them. In a charged capacitor with a DC voltage difference of V_B between its electrodes, the generated attractive force on the electrodes, F_e is given by:

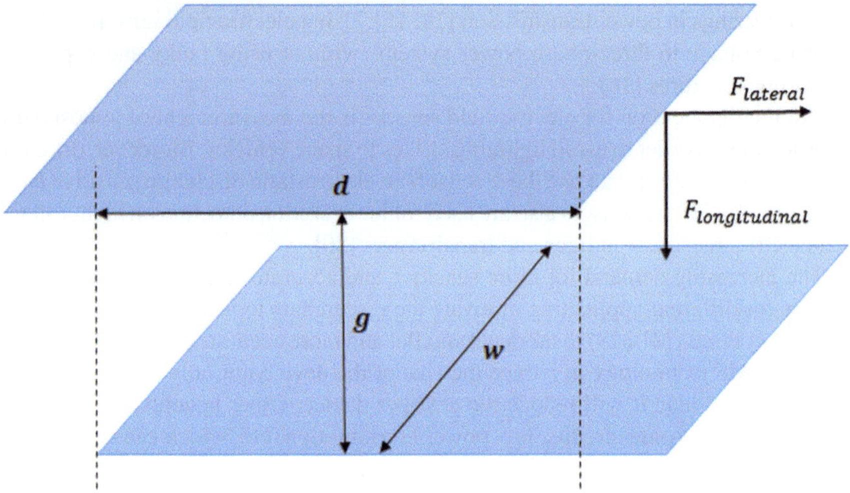

Fig. 1 A schematic of a capacitor with two parallel plates

$$F_e = \nabla W_e = \frac{1}{2}\nabla\left(CV_B^2\right) = \frac{V_B^2}{2}\nabla C \tag{1}$$

The electrodes can move laterally (parallel to each other) or longitudinally (orthogonal) with respect to each other. As it is presented in Fig. 1, the capacitance of a parallel plate capacitor which is also used in micromachined devices for actuation and sensing purposes is given by $C = \varepsilon\frac{A}{g}$ where ε is the permittivity of the dielectric, A is the effective overlap area between the plates, and g is the gap between them.

Lateral force ($F_{lateral}$) and longitudinal force ($F_{longitudinal}$) are given by:

$$F_{lateral} = \frac{\varepsilon w}{2g}V^2 \tag{2}$$

$$F_{longitudinal} = \frac{\varepsilon wd}{2(g-y)^2}V^2 \tag{3}$$

where w and d are the width and length of the part of the plates interacting with each other and y is the relative displacements of the electrodes from their initial gap. This is important to note that the gap between the electrodes is much smaller compared to the dimension of the plates. In order to produce larger force and also increase the sensitivity of the devices which are working based electrostatic actuation and capacitive sensing, one can increase the interactive are of the capacitor by using interdigitated fingers. Figure 2 shows a simple schematic of comb-drive consists of fingers. MEMS devices mainly use the comb-drive structure for electrostatic sensing [22] and actuation [23–25]. In actuation mode, the force generated between fingers is the source of the energy needed for vibration. In sensing mode, the change in the

Fig. 2 A regular comb-drive for increasing the force and sensitivity

capacitance due to the change in the gap or in the area can be detected and converted to the current or voltage as the output.

2.2 *Basic Resonator*

Resonators defined as the vibrating mechanical devices which are capable to convert kinetic and potential energy to each other using a spring mechanism. The response of the system depends on the frequency of the excitation. When the device is working in its optimum state which will happen in a certain frequency, the energy loss would be minimal. This certain frequency that causes the optimum situation known as the resonant frequency of the device. When the device is working on its resonant frequency, the response of the system shows a peak in this frequency. Moreover, for each vibrating system, there can be several resonant frequencies showing a specific pattern of vibration and movement known as the mode shapes of the system. Due to energy loss in every mechanical system, the response is highly related to the energy loss in that system. As the energy loss of the system decreases, the peak of the response gets sharper. This energy loss is referred to the damping of the system.

In order to model a resonator, one can use a simple mass-spring-damper system as shown in Fig. 3.

The equation of the motion of the system is given by:

$$M_{eff}\frac{\partial^2 x}{\partial t^2} + \zeta_{eff}\frac{\partial x}{\partial t} + K_{eff}x = F_{in} \tag{4}$$

where F_{in} is the input force, M_{eff} is the effective mass of the system, K_{eff} is the effective mechanical stiffness, and ζ_{eff} is the effective damping coefficient in the system. The input force in electrostatic devices referred to F_e in Eq. (1).

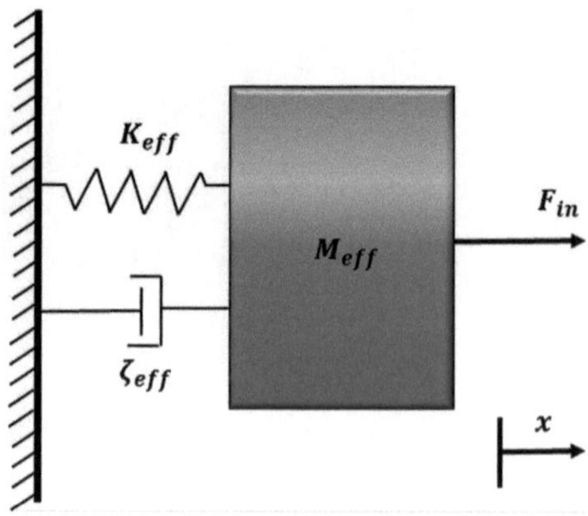

Fig. 3 Simple mass-spring-damper system to represent a resonator

The resonators have some important characteristics which will define the static and dynamic behavior of the system:

$$f_n = \frac{1}{2\pi}\sqrt{\frac{K_{eff}}{M_{eff}}} \tag{5}$$

$$\omega_n = 2\pi f_n \tag{6}$$

$$\xi = \frac{\omega}{\omega_n} = \frac{f}{f_n} \tag{7}$$

$$Q = \sqrt{\frac{M_{eff} K_{eff}}{\zeta_{eff}}} \tag{8}$$

$$\varphi = \tan^{-1}\frac{\xi}{Q(1-\xi^2)} \tag{9}$$

$$x_0 = \frac{F_{in}}{K_{eff}}\frac{1}{\sqrt{\left(1-\omega^2\right)^2 + (\xi/Q)^2}} \tag{10}$$

Which define the undamped resonant frequency of the system (or natural frequency), the radian frequency, the frequency ratio, the quality factor, the phase, and the static displacement respectively. The quality factor is of the significant characteristic of the system which represents the dynamic behavior of the system. Having a high Q factor is possible through minimizing the energy dissipation in the structure. The Q factor is formally defined as:

$$Q = 2\pi\frac{Average\, stored\, energy}{Energy\, lost\; per\; cycle} \tag{11}$$

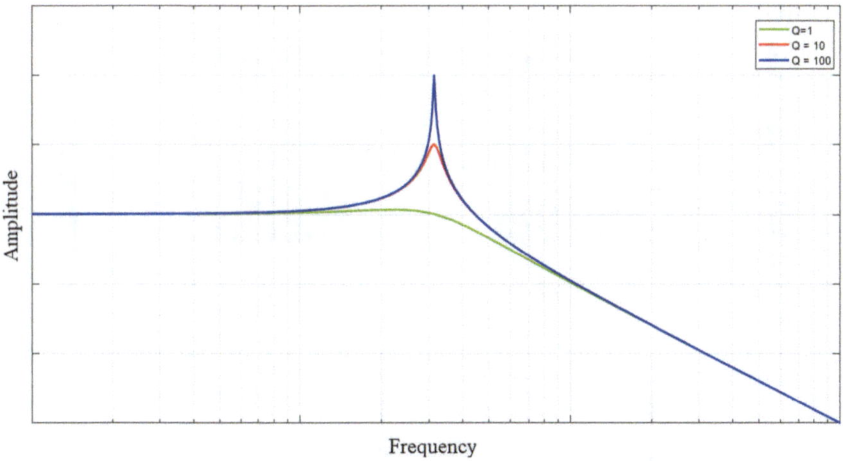

Fig. 4 The amplitude of vibration at resonant frequency depends on the quality factor. By increasing the Q, the resonant amplitude will increase proportionally

The relationship between the resonant frequency f_r and the undamped natural frequency f_n of a second order system is [26]:

$$f_r = f_n\sqrt{1 - \frac{1}{2Q^2}} \tag{12}$$

A high Q factor, as is common among MEMS resonators, makes the natural frequency and the resonant frequency of a device to be roughly equal. High-Q systems exhibit a peak in their amplitude at resonance (see Fig. 4).

2.3 Coupled Resonators

Single resonators can be coupled to each other and form an array of the identical resonator. The additional degrees of freedom from the coupling of the resonators to each other can be employed to enhance the sensitivity. As each resonator has the same mode shape, by coupling resonators once can increase the number of the mode shapes. In a coupled resonating system, resonators are connected to each other through different mechanisms of coupling such as mechanical, electrical or optical [27]. Figure 5a shows a schematic of a coupled resonator. The coupling between resonators (with the spring constant of K_c) can be either strong or weak. Strongly coupled resonators are mechanically coupled while weakly coupled resonators (WCRs) can be mechanically or electrostatically coupled. If the resonators are strongly coupled, the frequency separation between the modes gets widened (Fig. 5b). Weak mechanical coupling is possible by using a coupling structures with lengths that are odd multiples

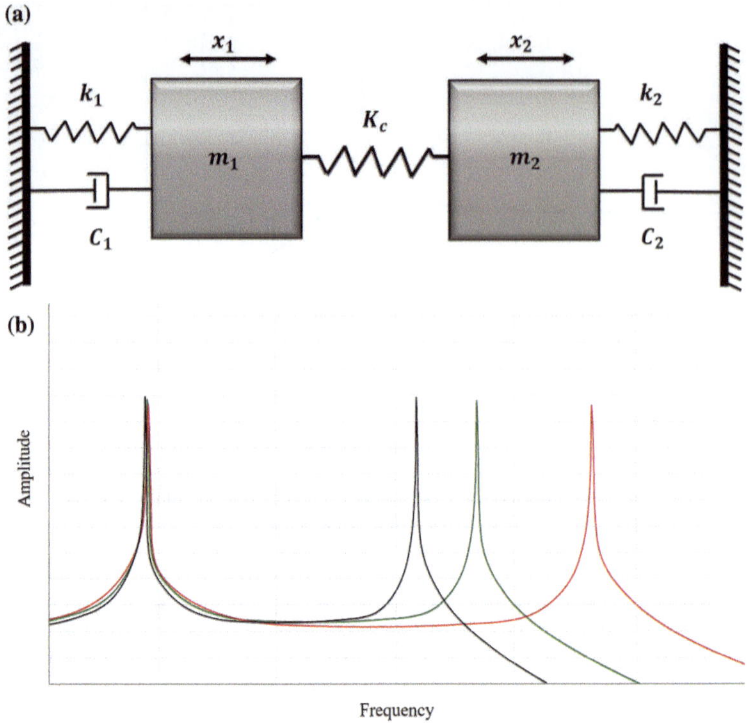

Fig. 5 **a** Simple model for a coupled resonator. **b** By increasing the coupling strength, the resonators responses are increasingly influenced by the other one, and hence, the frequency separation becomes wider

of a quarter of the acoustic wavelength [28]. Adding the coupling element between the resonators affects the frequencies and separates them from each other.

The main disadvantage of the mechanical coupling of the resonators is that it is not possible to tune the coupling between the resonators after the fabrication. Alternatively, the electrostatic spring which originates from the electrostatic force in the capacitor's gap can be used to realize the electrostatic coupling. Tuning the voltage across the capacitor's plates alters the coupling spring and as a result, it changes the separation between the modes [29]. Using Eq. (4), the dynamic equation of a 2DOF coupled resonator (Fig. 5) is given by:

$$\begin{bmatrix} m & 0 \\ 0 & m \end{bmatrix} \begin{bmatrix} \ddot{x}_1 \\ \ddot{x}_2 \end{bmatrix} + \begin{bmatrix} k + k_c & -k_c \\ -k_c & k + k_c + \Delta k \end{bmatrix} \begin{bmatrix} x_1 \\ x_2 \end{bmatrix} = 0 \tag{13}$$

In Fig. 5, the resonators are identical, so $m_1 = m_2 = m$ representing the mass for each resonator and $k_1 = k_2 = k$ which represents the spring constant of each resonator. Also k_c is the coupling stiffness and Δk is the stiffness perturbation. The eigenvalues and eigenvectors can be calculated as:

$$\omega_i^2 = \frac{2k + \Delta k + 2k_c \pm \sqrt{\Delta k^2 + 4k_c^2}}{2m} \tag{14}$$

$$u_i = \frac{x_{2i}}{x_{1i}} = \frac{\Delta k \pm \sqrt{\Delta k^2 + 4k_c^2}}{2k_c} \quad i = 1, 2 \tag{15}$$

where ω_i and u_i represents the resonant frequency and amplitude ratio of *ith* vibrating mode, respectively [30].

3 Micromachined Resonant Electrometers

3.1 Modulated Stiffness

Resonance sensing (or modulated stiffness) is one of the main methods used for sensing applications in resonator-based devices as it offers both high resolution and large dynamic. The resonant frequency of the device is a function of the measured. Any input to the system which causes a disturbance in the resonant element can alter the resonance frequency. This change can be through the stiffness, mass or geometry of the resonator.

These devices have two different mechanisms: the exciting mechanism and sensing mechanism. The first mechanism excites the movable part of the structure to vibrate in a specific direction. This excitation can be achieved by different mechanisms consist of electrostatic actuation, thermal actuation, piezoelectric actuation, etc. Since resonators are high-frequency systems, electrometers are mostly based on electrostatic actuation.

The sensing mechanism is responsible to sense this vibration in the form of frequency or mode shapes. This detected frequency can turn to the voltage as the output of the system. For electrometry, an applied charge to the system can cause an electrostatic force or a strain across the structure leading to the change in the frequency. As mentioned before in order to have a high-resolution sensor, it is necessary to have a high Q factor because the resolution is limited by the smallest detectable change in the resonant frequency of the resonator.

There are two popular MEMS-based resonators for electrometry. The first method is using an axial strain modulation or lateral stiffness perturbation of a flexural resonator, such as a double-ended tuning fork (DETF), to measure charge.

As shown in Fig. 6, the device is designed such that the electrostatic force from the input charge is applied axially to the resonator beam. The resonator beam is constrained horizontally, for instance, by a pair of suspension beams. The charge input electrode is located adjacent to the constrained end of the resonator and forms a couple of charge actuation capacitor. During the operation of the resonator, the gap between the input charge electrode and the constrained end function as a nonlinear electrical spring as a result of the produced electrostatic force. The input charge on the electrode will add an electrostatic strain which will change the effective stiffness

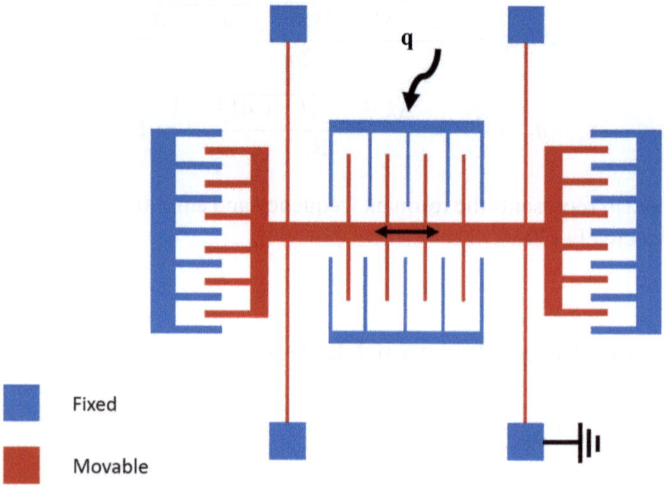

Fig. 6 Schematic of an electrometer vibrating by two comb drives at each side. The input charge on the upper part makes it possible to use differential sensing

of the resonator and shift the resonance frequency. The electrostatic force because of the input charge q is [28]

$$F_q = \frac{1}{2}\frac{\partial}{\partial y}\left(\frac{q^2}{C_q}\right) \tag{16}$$

$$F_q = \frac{q^2}{2\varepsilon H W_q}$$

where C_q is the capacitance of the charge sensing electrode, H is the height and W_q is the width of the electrode. The frequency shift due to the axial force in a DETF during the vibration represented by [28]

$$\Delta f = f_0\left(\frac{0.0375}{\varepsilon Y\, W_q W^3}\right)\left(\frac{L}{H}\right)^2 \tag{17}$$

where L is the length, W is the width of the DETF's beam and Y is Young's modulus of elasticity. Lee et al. reported an electrometer based on a DETF with the minimum measurable charge of 4 fC [31]. In other work, Chen et al. used the input charge to change the stiffness of the beams which formed DETF and are connecting to the drive combs. The reported sensitivity of this device is 4.4×10^{-4} Hz/fC^2 [32]. Zhao et al. used the DETF resonator for the charge sensing part and utilized an electrostatic comb-drive actuator to adjust the sensitivity. The electrostatic comb-drive is coupled to the charge input electrode using an adjustable capacitor for maximizing the sensitivity. The maximum sensitivity reported for the device is 1.03×10^{-2} Hz/fC^2 [33].

As the strain in the structure can induce the nonlinearity and amplitude-frequency (A-Z) effect which can cause frequency hysteresis and Duffing bifurcation instabilities, Chen et al. used a clamped-free semi-circular beams (CFCB) resonator. Using this structure, the fabricated electrometer had a sensitivity of 4.8×10^{-4} Hz/fC^2 [34].

3.2 Reed Electrometers

One of the common techniques for detecting electric charges in ambient condition and room temperature is using vibrating reed capacitor in a MEMS electrometer [6, 7, 35–40]. This method is based on using a balanced comb resonator with the ability to convert a DC charge input to an AC voltage output.

As Fig. 6 shows, an electrometer is created by differentially actuated resonating structures using comb drives on both sides of it. When the resonator is vibrating at its resonant frequency off, the capacitance between the fingers varying at a frequency of 2f. When a charge reaches the upper part of the stationary sensing fingers, it will induce a voltage at a frequency of 2f. This frequency is named the second harmonic (2nd harmonic) [41]. The reason that the charge is only inputted on the upper part is the possibility of differential sensing by subtracting the voltage signal of the bottom electrodes from the sense signal. The difference in frequency between the drive and sense voltage reduces feed-through voltage noise.

One of the advantages of sensing the output at the 2nd harmonic is to avoid the signal feedthrough from the comb electrodes used to actuate the device. Moreover, detecting the output at a higher frequency leads to an increase in the resolution since the device is working above the noise corner frequency of the circuit [41].

A schematic of a typical configuration for a resonator is depicted in Fig. 7. There are a pair of comb drives on both sides of a movable mass suspended by springs. The electrostatic force is used as the source of actuation. When an AC voltage with a DC bias is applied between the combs, the mass is vibrated in a direction parallel to the comb fingers.

The equation of the motion for this resonator is presented in Eq. (4). For the electrostatic actuation in this case, F_{in} is the electrostatic force between the fingers (F_e).

The electrostatic force on both sides of the mass are:

$$F_{el} = n\frac{\varepsilon t}{g}V_l^2 \tag{18}$$

$$F_{er} = n\frac{\varepsilon t}{g}V_r^2 \tag{19}$$

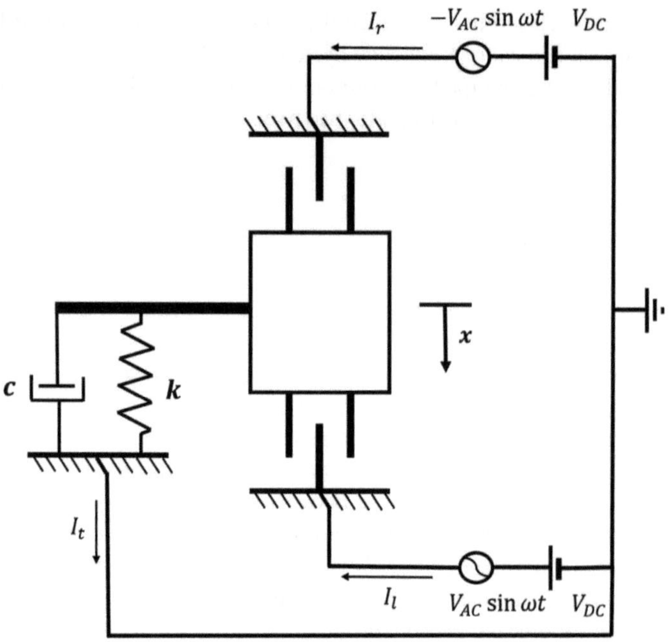

Fig. 7 Schematic of the configuration for an electrometer working based on capacitive sensing

where F_{el} and F_{er} are the electrostatic force on the left and right combs respectively, n is the number of the comb fingers on one side, ε is permittivity, t is the thickness of the fingers, g is gap between fingers, and V_r and V_l are the actuation voltage on right and left combs. By having $V_l = V_{DC} + V_{AC}\sin\omega t$ where ω is the angular velocity and V_r is equal but 180° out of phase the electric force will be:

$$\begin{aligned} F_e = F_{el} - F_{er} &= n\frac{\varepsilon t}{g}(V_{DC} + V_{AC}\sin\omega t)^2 - n\frac{\varepsilon t}{g}(V_{DC} + V_{AC}\sin\omega t)^2 \\ &= 4n\frac{\varepsilon t}{g}V_{AC}V_{DC}\sin\omega t \\ &= F_0\sin\omega t \end{aligned} \tag{20}$$

$$F_0 = 4n\frac{\varepsilon t}{g}V_{AC}V_{DC} \tag{21}$$

where, by substituting Eqs. (4) and (20) and solving the linear solution, x will be:

$$x = x_0\sin(\omega t - \varphi) \tag{22}$$

where x_0 and φ are defined in Eqs. (9) and (10).

Exciting the resonator is through the vibration mechanism while the sensing mechanism detects this vibration in the form of frequency. The frequency of the applied vibration affects the output of the sensor in the form of voltage.

Considering l as the initial overlap of the fingers, the capacitance of the left and right combs is different and given by [42]:

$$C_l = 2n\frac{\varepsilon t(l-x)}{g}$$
$$C_r = 2n\frac{\varepsilon t(l+x)}{g} \tag{23}$$

And the produced charge on each side are:

$$q_l = C_l V_l$$
$$q_r = C_r V_r \tag{24}$$

The induced current is charges change with respect to the time

$$I_l = V_l\frac{dC_r}{dx}\dot{x} + C_l\frac{dV_r}{dt} = V_l\frac{dC_l}{dx}\dot{x} + C_r\frac{dV_l}{dt} \tag{25}$$

$$I_r = V_r\frac{dC_r}{dx}\dot{x} + C_r\frac{dV_r}{dt} = -V_r\frac{dC_l}{dx}\dot{x} - C_r\frac{dV_l}{dt} \tag{26}$$

where $\dot{x}$ is the velocity of the mass.

The total current is calculated by adding Eqs. (24) and (25):

$$I_t = I_r + I_l = (V_l - V_r)\frac{dC_l}{dx}\dot{x} + (C_l - C_r)\frac{dV_l}{dt}$$
$$= -\frac{4n\varepsilon t}{g}V_{AC}\,x_0\,\omega\sin(2\omega t - \phi) \tag{27}$$

Using Eqs. (10), (22) and (27) gives a signal at twice the actuation voltage frequency:

$$I_t = -\left(\frac{4n\varepsilon t}{g}\right)^2\frac{V_{AC}^2 V_{DC}}{k}\omega_n\frac{1}{\sqrt{\left(1-\omega^2\right)^2 + (\xi/Q)^2}}\sin(2\omega t - \phi) \tag{28}$$

The induced voltage is equal to [21]:

$$V(t) = \frac{q}{C_v(t) + C_p} \tag{29}$$

where q is the input charge, $C_v(t)$ denotes the variable capacitance of the fingers, and C_p is the parasitic capacitance to the ground. The RMS magnitude of this second harmonic component of the output voltage varies linearly with the input charge and is given by [21]:

$$V(2f) = q\frac{C_s}{2\sqrt{2}(C_v + C_p)^2}\left(\frac{X}{g}\right)^2 \tag{30}$$

where C_s the static capacitance of the variable capacitor, X is the amplitude of the resonator displacement, and g is the initial gap between the fingers. By getting the derivation from the output RMS voltage with respect to the input charge the resolution of the electrometer will be calculated as it can be seen in Eq. (31). To have the maximum charge resolution for a given X and g, C_p should be equal to C_s [37].

$$R_e = \frac{C_s}{2\sqrt{2}(C_v + C_p)^2}\left(\frac{X}{g}\right)^2 \tag{31}$$

Riehl et al. reported the first electrometer and electrostatic field meter using vibrating reed technique. Their device had a charge resolution of 51 $e/\sqrt{Hz}$ [12]. Using this method, Lee et al. [41] fabricated a device which has an equivalent noise floor of 245 $e/\sqrt{Hz}$ at room temperature and atmospheric pressure, and a voltage-to-charge gain of 1.02×10^8 V/C. Later, in other works the resolution in the sub-10 electron regime was demonstrated by [37–39].

3.3 Coupled Electrometers

In capacitance sensing, where the input charge is employed to modify the value of the device's variable capacitance, the sensor response is susceptible to interference from the drive signals. Moreover, in resonance sensing where the input charge induces stress in structure and shifts the resonant frequency, there are some differences between identical devices due to inherent tolerance in microfabrication which leads to a difference in their sensitivity to interfering parameters, such as temperature variations [43]. In order to address these issues and improve the sensitivity of the device, it is possible to utilize the WCRs. The stiffness disorder of the WCRs caused by the fabrication tolerances can be compensated by independent electrostatic tune electrodes [44]. The coupled resonators make it possible to exploit differential signal as the output and the electrometer can work based on the difference in resonance frequencies which will cancel the effect of fabrication and environmental disturbances [45].

One of the common techniques that has been used in WCRs is mode localization. Mode localization involves the manipulation of energy between two coupled modes and is one of the main methods used in coupled resonator-based electrometers [44]. This technique is based on the fact that each mode of resonance is sensitive to the perturbations to one of the coupled resonators. This can result in providing a

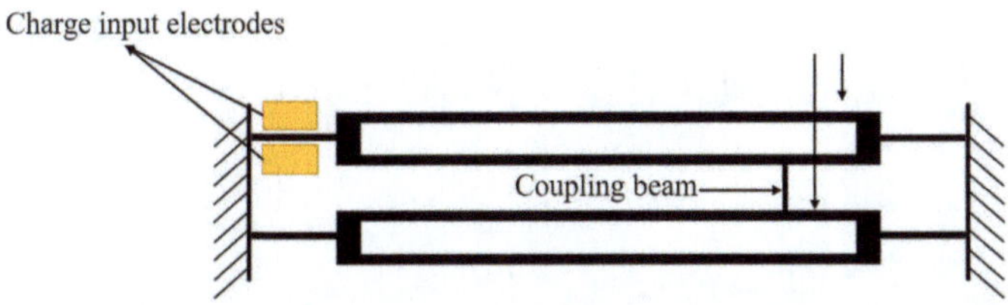

Fig. 8 A schematic of a 2DOF WCR consists of two identical DETF and two charge input electrodes

specific response to characterize the pattern of the mode shapes of the system. In the mechanical vibration, the mode localization phenomenon can happen when there is a small symmetry-breaking perturbation on the nearly symmetrical weakly coupled resonator structure. The presence of this small perturbation will result in a vibration energy confinement and a significant change in the mode shape [46]. Using the mode shape/amplitude ratio-based sensitivity instead of the frequency, the sensitivity will be two or three orders of magnitude higher [30]. Figure 8 shows a schematic of a 2DOF WCR consist of two identical DETF and two charge input electrodes sandwiched the resonator 1 spring.

Applying an external charge to the input charge electrode will cause an electrostatic force between the input electrode and the resonator 1's spring. This force will introduce an axial strain to the spring beam and changes the effective stiffness of resonator 1. Since the resonator 2 will remain in the initial condition the symmetry of the system will break, and mode localization will occur. Therefore, the amplitude of the two resonators is not equal anymore. It is worth to mention that the sensitivity will be significantly amplified if the amplitude ratio is used as the output of the system. By monitoring the variations of the amplitude ratio of the WCRs, the quantities of the input charge can be measured.

The sensitivity will be significantly amplified if the amplitude ratio is used as the output metric. By measuring the variations of the amplitude ratios of the WCRs, the quantities of the input charge can be obtained.

Zhang et al. [30] used the mode localization in a 2DOF WCRs (Fig. 9) and reported that the amplitude ratio-based sensitivity is around 215 times higher than the frequency-based sensitivity. The reported amplitude ratio-based resolution of the electrometer was 1.269 fC.

Moreover, Yang et al. [46] used a 3DOF weakly coupled resonators and reported the resolution of 9.21 $e/\sqrt{Hz}$. In another research [47], a 4DOF WCRs a 4DOF mode-localized electrometer with 0.256 $e/\sqrt{Hz}$ charge resolution is reported, which is the highest level of micromachined electrometers operating at room temperature up to now.

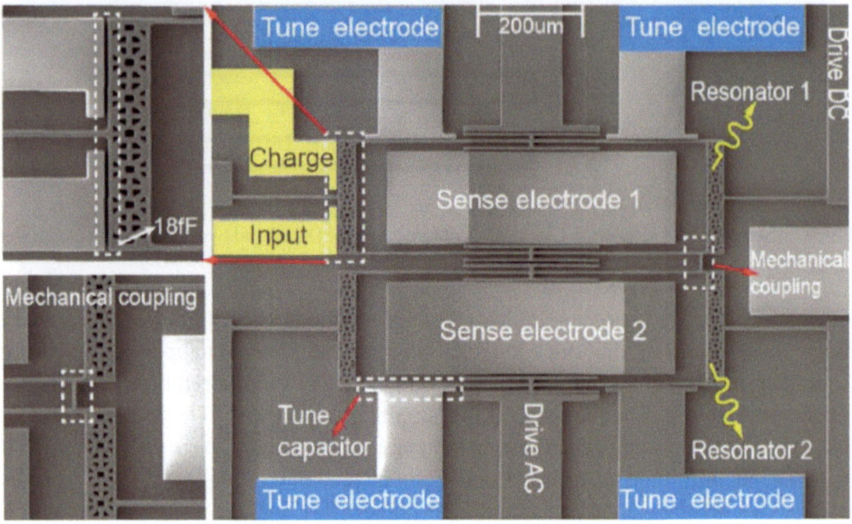

Fig. 9 An SEM photo of 2DOF weakly coupled resonator fabricated by [30]

3.4 Electric Field Sensors

Electric field sensors are field measuring devices which work on the electrostatic induction principle. When a conducting plate is in effective electrical contact with the ground and exposed to an electric field, it will collect a net charge on the surface. The amount of induced charges can be calculated using Gauss's law:

$$q = \varepsilon_r \varepsilon_0 E A \tag{32}$$

where A is the area, q is the amount of induced charges by the electric field E. If the exposed electrode is covered from the incident electric field, charges on the surface would go back to the ground and when the electrode is again exposed to the electric field, charges will again travel to the electrode. By covering and exposing the sensing electrodes periodically, an alternate current can be generated between exposed and covered parts of the electrodes and it is proportional to the field intensity, shutter working frequency and the sensing electrodes area. The magnitude of this induced current is given by Eq. (33). By measuring the current, the electric field sensor can measure the electric field intensity.

$$i = \frac{dq}{dt} = \varepsilon_r \varepsilon_0 E \frac{dA}{dt} \tag{33}$$

Due to the periodic variation of the AC electric field with respect to time, it is relatively easy to measure. On the opposite side, when a probe is exposed to a DC field, it will collect the charges gradually and after a while, all parts of the probe will

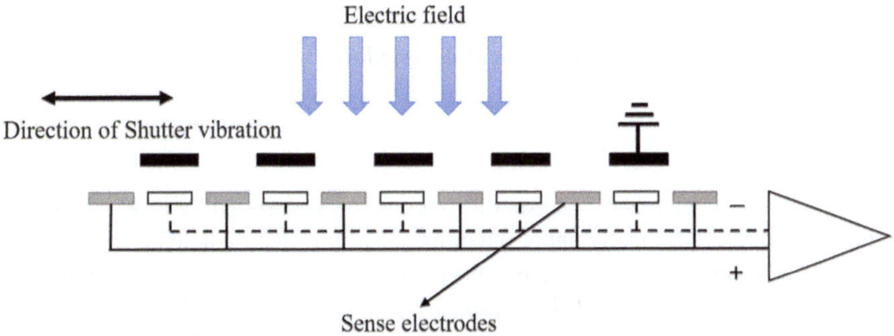

Fig. 10 Schematic view of the shutter and the sensing electrode

have equal potential. Therefore, the output of the probe will diminish and reach zero. Because of the mentioned reason, there should be a mechanism to convert the DC field to an AC field by shielding and covering the sensor periodically.

In most micromachined electric field sensors (MEFS), the structure consists of a moveable shutter above a sensing part. Vibrational movement of the shutter leads to shield and uncover the sensing electrodes [21, 48, 49]. Figure 10 presents a schematic view of the shutter and the sensing electrode.

MEMS electric field sensors can be categorized according to their working principle, including the mechanical mechanism for the shutter. In a mechanical vibrating mechanism, most of the time there is a vibrational mechanical shutter in order to turn the DC electrostatic field to AC output current. The sensing part can be considered as a proof mass attached to the mainframe through the number of springs. There are few driving mechanisms for sensing and measuring the electric field intensity. These mechanisms mainly include actuators using electrostatic mechanism, thermal driven method [50–52], and piezoelectric mechanism [53, 54].

Moreover, MEFS can be divided into two categories according to the shutter movement direction including lateral movement and vertical movement. There are several mechanisms for using electrostatic force for electric field sensing. In one structure, it consists of electrostatic-driving electrodes, sensing combs, shutter combs, and spring beams. By applying a bias voltage between electrode fingers and shuttle fingers, the movable shuttle will start to vibrate. The driving voltage consists of a DC and an AC voltage leads the shutter to vibrate at the frequency of the AC voltage. The shutter in this mechanism can vibrate in two different ways:

1. It can be placed over the sensing electrodes to cover and uncover the sensing area and as a result, expose and shield the sensing area to the electric field. The shutter vibration will expose and cover the sensing electrode, leads the charges accumulation and depletion on the electrode surface and generating an AC current.
2. Shutter and the sensing part are placed in the same plane [53, 55, 56]. The shutter consists of shielding electrodes which can vibrate horizontally. When the shielding electrodes are near to the sensing part, the shielding electrode will

prevent the electric field to reach the region around the sensing electrode and reduces the inducted charge as it is depicted in Fig. 11. By moving the shutter away from the sensing electrode, the electric field can reach the region around the sensing electrode, increasing the inducted charge. By detecting this inducted charge in the sensing electrode, it is possible to measure the electric field.

The second mechanism consists of a membrane suspended by a set of springs, which is displaced vertically under the influence of the electric field being measured (Fig. 12). The vertical displacement of the membrane (Δd) is indicative of the electric field strength and is measured either capacitively or by a laser employed system. The voltage difference between the upper electrode and the sensing electrode (metalized membrane) creates an electric field force that pulls the sensing electrode upwards.

Chen [57] presented an electric field using the electrostatic force principle. In his work, the electric field strength is measured capacitively. This system consists of two electrodes. The spring-suspended membrane is connected to the ground and the bottom electrode is at floating potential creating a voltage difference between the two

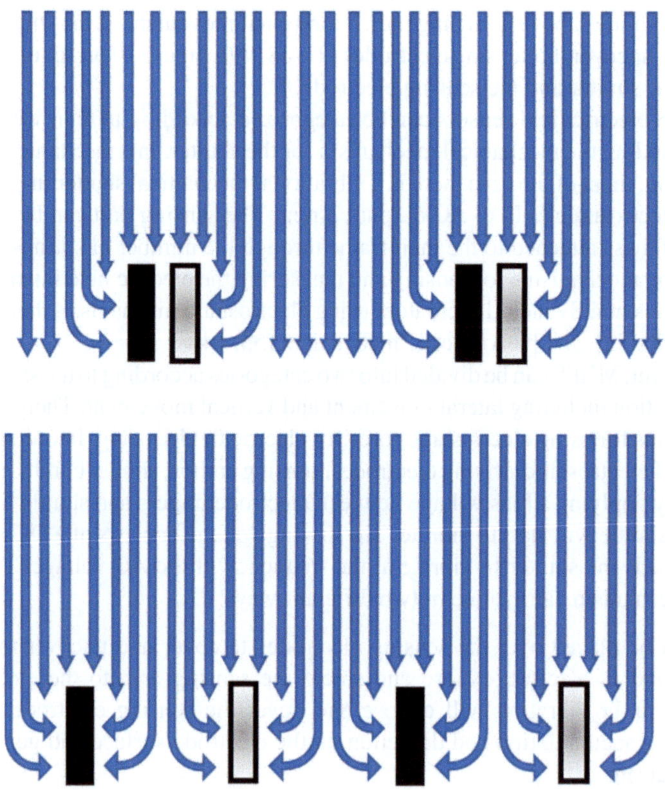

Fig. 11 The shutter moving in the same plane with the sensing electrode, preventing the electric field to go around the sensing electrode fingers

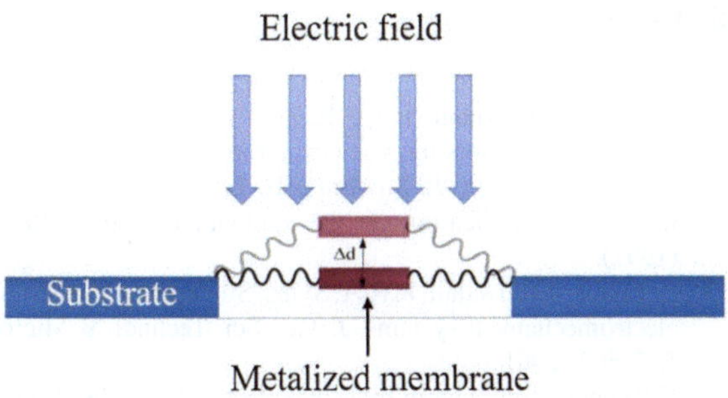

Fig. 12 Electrostatic force due to the electric field displaces the metalized membrane vertically

electrodes. Electric field force pulls the membrane towards the power source changing the distance between the two electrodes. This creates a change in the capacitance which is the factor of the measurand.

Figure 13 shows a schematic electrostatic force-based sensor monitored by a laser interrogated the system. When an electric field is present, the membrane mirror lifts, and beam position of the reflected laser beam moves into a different position on the surface of the photodetector. The laser displacement is measured electronically and the difference between the output currents of photodetectors is used as the sensing signal to determine the strength of the electric field.

Roncin et al. [58] used an electrostatic force deflected membrane supported by eight spring-beams. They reported a resolution of 5 kV/m under a dc field and a 0.3 V/m resolution under an ac field at 97 Hz. Chen et al. [59] reported a MEFS with a resolution of 17 V/m and a sensitivity of 360 mV/(kV/m).

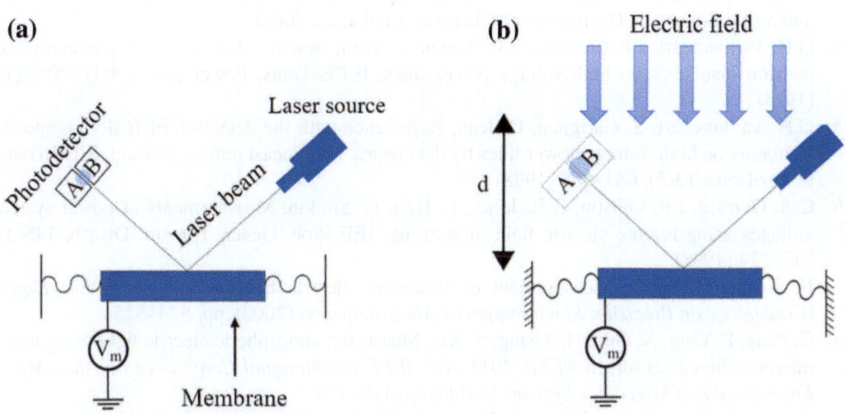

Fig. 13 The vertical displacement of the membrane due to the electric field is sensed using a laser beam reflected from the surface, **a** without an electric field, **b** with an electric field

References

1. C.I. Calle, J.G. Mantovani, C.R. Buhler, E.E. Groop, M.G. Buehler, A.W. Nowicki, Embedded electrostatic sensors for Mars exploration missions. J. Electrostat. **61**(3–4), 245–257 (2004)
2. R. Heer, C. Eder, J. Smoliner, E. Gornik, Floating electrometer for scanning tunneling microscope applications in the femtoampere range. Rev. Sci. Instrum. **68**(12), 4488–4491 (1997)
3. S. Taylor, R.F. Tindall, R.R.A. Syms, Silicon based quadrupole mass spectrometry using micro-electromechanical systems. J. Vac. Sci. Technol. B Microelectron. Nanometer Struct. **19**(2), 557–562 (2001)
4. F. Ruggeri et al., Single-molecule electrometry. Nat. Nanotechnol. **12**(5), 488–495 (2017)
5. F. Krueger, J. Larson, Chipmunk IV: development of and experience with a new generation of radiation area monitors for accelerator applications. Nucl. Instrum. Methods Phys. Res. A Accel. Spectrometers Detect. Assoc. Equip. **495**(1), 20–28 (2002)
6. G. Jaramillo, M. Li, C. Buffa, F.J. Brechtel, D.A. Horsley, Charged particle detection using a micromechanical electrometer, in *Technical Digest—Solid-State Sensors, Actuators, and Microsystems Workshop* (2012), pp. 295–298
7. G. Jaramillo, C. Buffa, M. Li, F.J. Brechtel, G. Langfelder, D.A. Horsley, MEMS electrometer with femtoampere resolution for aerosol particulate measurements. IEEE Sens. J. **13**(8), 2993–3000 (2013)
8. J. Jalil, Y. Zhu, C. Ekanayake, Y. Ruan, Sensing of single electrons using micro and nano technologies: a review. Nanotechnology **28**(14) (2017). Institute of Physics Publishing
9. N. Clement, H. Inokawa, Foundry metal-oxide-semiconductor field-effect-transistor electrometer for single-electron detection. Jpn. J. Appl. Phys. Part 1 Regul. Pap. Short Notes Rev. Pap. **44**(7A), 4855–4858 (2005)
10. *Low Level Measurements Handbook*, 7th ed. (Keithley, 2013)
11. R. Pallás-Areny, J.G. Webster, *Sensors and Signal Conditioning* (Wiley, 2001)
12. K. Nishiguchi, Y. Ono, A. Fujiwara, Single-electron thermal noise. Nanotechnology **25**(27) (2014)
13. BP International and Institution of Chemical Engineers (Great Britain), *Hazards of Electricity and Static Electricity* (Institution of Chemical Engineers, 2006)
14. G.H. Vaillancourt, J.P. Bellerive, M. St-Jean, C. Jean, New live line tester for porcelain suspension insulators on high-voltage power lines. IEEE Trans. Power Deliv. **9**(1), 208–219 (1994)
15. G.H. Vaillancourt, S. Carignan, C. Jean, Experience with the detection of faulty composite insulators on high-voltage power lines by the electric field measurement method. IEEE Trans. Power Deliv. **13**(2), 661–666 (1998)
16. C.A. Gerrard, J.R. Gibson, G.R. Jones, L. Holt, D. Simkin, Measurements of power system voltages using remote electric field monitoring. IEE Proc. Gener. Transm. Distrib. **145**(3), 217–224 (1998)
17. H. Kirkham, On the measurement of stationary electric fields in air, in *CPEM Digest (Conference on Precision Electromagnetic Measurements)* (2002), pp. 524–525
18. C. Peng, P. Yang, X. Guo, H. Zhang, S. Xia, Measuring atmospheric electric field using novel micromachined sensor, in *NEMS 2011—6th IEEE International Conference on Nano/Micro Engineered and Molecular Systems* (2011), pp. 417–420
19. C. Peng, X. Chen, Q. Bai, L. Luo, S. Xia, A novel high performance micromechanical resonant electrostatic field sensor used in atmospheric electric field detection, in *Proceedings of the IEEE International Conference on Micro Electro Mechanical Systems (MEMS)*, vol. 2006 (2006), pp. 698–701
20. C. Barthod, M. Prasad, J. Bouillot, C. Galez, M. Farzaneh, High electric field measurement and ice detection using a safe probe near power installations. Sens. Actuator A Phys. **113**(2), 140–146 (2004)

21. P.S. Riehl, K.L. Scott, R.S. Muller, R.T. Howe, J.A. Yasaitis, Electrostatic charge and field sensors based on micromechanical resonators. J. Microelectromech. Syst. **12**(5), 577–589 (2003)
22. N. Yazdi, F. Ayazi, K. Najafi, Micromachined inertial sensors. Proc. IEEE **86**(8), 1640–1658 (1998)
23. M.A. Rosa, S. Dimitrijev, H.B. Harrison, Enhanced electrostatic force generation capability of angled comb finger design used in electrostatic comb-drive actuators. Electron. Lett. **34**(18), 1787–1788 (1998)
24. D.A. Horsley, N. Wongkomet, R. Horowitz, A.P. Pisano, Precision positioning using a microfabricated electrostatic actuator. IEEE Trans. Magn. **35**(2), PART 1, 993–999 (1999)
25. R. Legtenberg, A.W. Groeneveld, M. Elwenspoek, Comb-drive actuators for large displacements. J. Micromech. Microeng. **6**(3), 320–329 (1996)
26. B. Bahreyni, Chapter 4—Modelling of statics, in *Fabrication and Design of Resonant Microdevices*, ed. by B. Bahreyni (William Andrew Publishing, 2009), pp. 69–78
27. S. Gröblacher, K. Hammerer, M.R. Vanner, M. Aspelmeyer, Observation of strong coupling between a micromechanical resonator and an optical cavity field. Nature **460**(7256), 724–727 (2009)
28. S.S. Li, Y.W. Lin, Z. Ren, C.T.C. Nguyen, An MSI micromechanical differential disk-array filter, in *TRANSDUCERS and EUROSENSORS'07—4th International Conference on Solid-State Sensors, Actuators and Microsystems* (2007), pp. 307–311
29. R. Abdolvand, B. Bahreyni, J. Lee, F. Nabki, Micromachined resonators: a review. Micromachines **7**(9), 160 (2016)
30. H. Zhang, J. Huang, W. Yuan, H. Chang, A high-sensitivity micromechanical electrometer based on mode localization of two degree-of-freedom weakly coupled resonators. J. Microelectromech. Syst. **25**(5), 937–946 (2016)
31. J.E.Y. Lee, B. Bahreyni, A.A. Seshia, An axial strain modulated double-ended tuning fork electrometer. Sens. Actuator A Phys. **148**(2), 395–400 (2008)
32. D. Chen, J. Zhao, Y. Wang, J. Xie, Electrostatic charge sensor based on micro resonator with sensing scheme of effective stiffness perturbation, in *Proceedings of the IEEE International Conference on Micro Electro Mechanical Systems (MEMS)* (2017), pp. 1208–1211
33. J. Zhao, H. Ding, S. Ni, L. Fu, W. Wang, J. Xie, High-resolution and large dynamic range electrometer with adjustable sensitivity based on micro resonator and electrostatic actuator, in *Proceedings of the IEEE International Conference on Micro Electro Mechanical Systems (MEMS)*, vol. 2016, Feb 2016, pp. 1074–1077
34. D. Chen et al., High sensitivity micro electrometer based on clamped free curved beams resonator with weakened nonlinearity, in *Proceedings of the IEEE International Conference on Micro Electro Mechanical Systems (MEMS)*, vol. 2018, Jan 2018, pp. 1092–1095
35. P.S. Riehl, K.L. Scott, R.S. Muller, R.T. Howe, High-resolution electrometer with micromechanical variable capacitor (2002)
36. G. Jaramillo, D.A. Horsley, C. Buffa, G. Langfelder, A MEMS based electrometer with a low-noise switched reset amplifier for charge measurement, in *Proceedings of IEEE Sensors* (2012)
37. J.E.Y. Lee, Y. Zhu, A.A. Seshia, A micromechanical electrometer approaching single-electron charge resolution at room temperature, in *Proceedings of the IEEE International Conference on Micro Electro Mechanical Systems (MEMS)* (2008), pp. 948–951
38. J. Lee, Y. Zhu, A. Seshia, Room temperature electrometry with SUB-10 electron charge resolution. J. Micromech. Microeng. **18**(2) (2008)
39. J. Lee, Y. Zhu, A. Seshia, Sub-10e charge resolution for room temperature electrometry, in *Proceedings of IEEE Sensors* (2007), pp. 821–824
40. A. Menzel, A.T.H. Lin, P. Estrela, P. Li, A.A. Seshia, Biomolecular and electrochemical charge detection by a micromechanical electrometer. Sens. Actuator B Chem. **160**(1), 301–305 (2011)
41. J. Lee, Y. Zhu, A.A. Seshia, A variable capacitor based MEMS electrometer, in *Proceedings Eurosensors XX* (2006), pp. 452–453

42. G.C. Underwood, A MEMS dual vertical electrometer and electric field-mill. Recommended citation. AFIT scholar theses and dissertations student graduate works (2019)
43. M.S. Hajhashemi, B. Bahreyni, A differential electrometer based on coupled microresonators, in *Proceedings of IEEE Sensors* (2012)
44. H. Zhang, W. Yuan, J. Huang, B. Li, H. Chang, A high-sensitive resonant electrometer based on mode localization of the weakly coupled resonators, in *Proceedings of the IEEE International Conference on Micro Electro Mechanical Systems (MEMS)*, vol. 2016, Feb 2016, pp. 87–90
45. M.S. Hajhashemi, B. Bahreyni, Characterization of disturbances in systems of coupled microresonator arrays. IEEE Sens. J. **12**(7), 2510–2516 (2012)
46. J. Yang, H. Kang, H. Chang, A micro resonant electrometer with 9-electron charge resolution in room temperature, in *Proceedings of the IEEE International Conference on Micro Electro Mechanical Systems (MEMS)*, vol. 2018, Jan 2018, pp. 67–70
47. H. Kang, B. Ruan, Y. Hao, H. Chang, A micromachined electrometer with room temperature resolution of 0.256 e/$\sqrt{}$Hz. IEEE Sens. J. 1–1 (2019)
48. G. Wijeweera, B. Bahreyni, C. Shafai, A. Rajapakse, D. Swatek, Micromachined electric field sensor to measure AC and DC fields in power systems, in *2009 IEEE Power & Energy Society General Meeting* (2009), pp. 1–1
49. X. Chen et al., Thermally driven micro-electrostatic fieldmeter. Sens. Actuator A Phys. **132**(2), 677–682 (2006)
50. L. Que, J.S. Park, Y.B. Gianchandani, Bent-beam electrothermal actuators—part I: single beam and cascaded devices. J. Microelectromech. Syst. **10**(2), 247–254 (2001)
51. L. Que, J.S. Park, Y.B. Gianchandani, Bent-beam electro-thermal actuators for high force applications, in *Proceedings of the IEEE Micro Electro Mechanical Systems (MEMS)* (1999), pp. 31–36
52. B. Bahreyni, G. Wijeweera, C. Shafai, A. Rajapakse, Analysis and design of a micromachined electric-field sensor. J. Microelectromech. Syst. **17**(1), 31–36 (2008)
53. S. Ghionea, G. Smith, J. Pulskamp, S. Bedair, C. Meyer, D. Hull, MEMS electric-field sensor with lead zirconate titanate (PZT)-actuated electrodes, in *Proceedings of IEEE Sensors* (2013)
54. J. Huang, X. Wu, X. Wang, X. Yan, L. Lin, A novel high-sensitivity electrostatic biased electric field sensor. J. Micromech. Microeng. **25**(9) (2015)
55. T. Chen, C. Shafai, A. Rajapakse, B.Y. Park, Micromachined electric field mill employing a vertical moving shutter. Procedia Eng. **87**, 452–455 (2014)
56. P. Yang, C. Peng, H. Zhang, S. Liu, D. Fang, S. Xia, A high sensitivity SOI electric-field sensor with novel comb-shaped microelectrodes, in *2011 16th International Solid-State Sensors, Actuators and Microsystems Conference, TRANSDUCERS'11* (2011), pp. 1034–1037
57. T. Chen, MEM electric field sensor using force deflection with capacitance interrogation, in *IEEE Power and Energy Society General Meeting* (2013)
58. A. Roncin, C. Shafai, D.R. Swatek, Electric field sensor using electrostatic force deflection of a micro-spring supported membrane. Sens. Actuator A Phys. **123–124**, 179–184 (2005)
59. T. Chen, C. Shafai, A. Rajapakse, J.S.H. Liyanage, T.D. Neusitzer, Micromachined ac/dc electric field sensor with modulated sensitivity. Sens. Actuator A Phys. **245**, 76–84 (2016)

System-Level Modelling of MEMS Vibrating-Reed Electrometer in Matlab Simulink

Yong Zhu and Y. Kuang

Abstract Micromachined mechanical variable capacitor is a key component of many microelectromechanical systems (MEMS) devices, such as pressure sensor, accelerometer, gyroscope, electrometer, etc. Optimization of these devices require a systematic consideration of parameter design in both mechanical and electrical domains. This chapter introduces the system-level modelling of a variable capacitor-based MEMS electrometry system. To simultaneously simulate the mechanical and electrical components, Matlab Simulink was adopted as the simulation environment to build the system model. Four main modelling blocks were developed to model the entire system, including electrostatic force generator, equivalent circuit representation of spring-mass-damper mechanical system, time-dependent variable capacitor controlled by in-line equation, and integration of variable capacitor into the actual detection circuitry. The whole electrometry system was successfully simulated in Simulink, and the detailed simulation results are shown in this chapter, such as the waveforms of driving force, shuttle displacement, time-varying capacitance and charge induced output voltage. The simulated results agreed well with the experimental results, for example, the simulated charge sensitivity is 7.2×10^7 V/C, which is close to the experimental results of 9.5×10^7 V/C with the same design parameters. Using this technique, any variable capacitor-based MEMS sensors can be modelled in Simulink by simply changing the mathematical calculation blocks to implement the in-line equation for the time-dependent capacitance. Therefore, this technique provides a useful tool that allows a fully combined simulation of both mechanical and electronic systems.

Y. Zhu (✉) · Y. Kuang
School of Engineering and Built Environment, Griffith University, Gold Coast Campus, Gold Coast, QLD 4222, Australia
e-mail: y.zhu@griffith.edu.au

Y. Kuang
e-mail: y.kuang@griffith.edu.au

Y. Zhu (ed.), *Micro and Nano Machined Electrometers*,
https://doi.org/10.1007/978-981-13-3247-0_7

1 Introduction

Microelectromechanical Systems (MEMS) device, as its name indicates, has both mechanical structures and electronic circuits in operation [1, 2]. To obtain optimal device performance, one needs to design the device at a higher level through systematic approaches. Design parameters in both mechanical and electrical domains must be considered simultaneously, and sometimes compromised to each other. Therefore, in the design phase, a simulation technique is needed to fully combine mechanical and electrical components in the same simulation environment. This technique is required to aid the system-level performance modelling and the optimization of the design of both electromechanical structures and detection electronics.

As electronic circuit simulators, for example PSpice and Matlab Simulink, have been well developed and established as the industrial standard simulation environment for electronic circuit development. Therefore, it is convenient to adopt these circuit simulators as the system-level modelling tools for MEMS devices. To embed mechanical structures into electronic circuit simulators, the equivalent circuit representation of mechanical structures has been developed for both capacitive micromechanical accelerometers [3] and gyroscopes [4]. However, in general, the overall MEMS detection systems, including micro-structures and interface electronics, cannot be entirely implemented in a user-friendly simulation environment simultaneously. The reason is that the in-line equations for time-varying components are not available in most circuit simulators [5]. The equivalences between mechanical and electrical quantities can be used to build circuit models in electronic circuit simulator such as PSpice. However, most circuit simulators can only deal with linear or fix-value device simulations.

To implement the system-level modelling of a vibrating-reed electrometer, a nonlinear time-domain variable capacitor has been built in PSpice [6, 7]. In this model, a complicated PSpice simulation technique was proposed to model a variable capacitor by a voltage-controlled voltage source, whose value is controlled by an auxiliary integrating circuit and in-line equations. The technique was also implemented to model the nonlinearity in MEMS resonators by using a charge-controlled capacitor in PSpice [8, 9].

In this chapter, a more convenient technique to model the arbitrary nonlinear capacitors will be proposed and discussed using Simulink in Matlab. The simulated results obtained are found to be in a good agreement with experimental results. Using this method, any variable capacitor-based sensor can be simulated in Simulink, providing a powerful tool that allows a full combined simulation of both mechanical and electronic systems for studying MEMS.

The remainder of this chapter is organized as follows. Section 2 provides the background information of vibrating-reed electrometer. Section 3 is the main section of this chapter including the Simulink modelling of electrostatic force, MEMS resonator, time-varying capacitor, and whole mechanical-electrical systems. The simulation results are also shown in this section. Section 4 concludes this chapter.

2 MEMS Vibrating-Reed Electrometer Principles

An electrometer or charge sensor is an instrument used to measure weak electronic charge signals and have found extensive applications, such as detection of bio-analyte, tunnelling microscopy, mass spectrometry, ionization current detection in radiation chambers, aerosol science, and quantum computing [10–15]. In general, the electrometer's accuracy is limited by various factors, such as heterogeneity and substrate differences in sensor manufacturing, excessive inherent noises from sensing elements and electronic circuitry, external disturbances, etc. Charge resolution and sensitivity are the two key parameters to designate the property of electrometry systems. For example, a commercially available electrometer Keysight B2987A [16] has a resolution of 6,300 electrons at room temperature and ambient pressure, which is not accurate enough for high-end charge sensing scenarios. By contrast, a fraction of electrons can be detected by single electron transistors (SETs) [17], nano-capacitive gate sensing [18] and nanoelectromechanical systems (NEMS) resonator-based electrometers [14], however stringent conditions such as cryogenic temperature, vacuum pressure and non-standard fabrication processes are required in practical applications.

The main reason for poor resolution of commercial electrometers (for example, Keysight B2987A) is the use of noisy DC preamplifiers. As the charge in real-life applications is mainly DC or slowly changed quantity, the charge signal can be easily immersed in the high magnitude of flicker noise or 1/f noise (near DC levels or low frequencies) at the input terminal of preamplifier. To solve this problem, the vibrating-reed electrometer was invented to up-modulate the input DC charge to a higher frequency above the corner frequency of flicker noise [20–24]. A vibrating-reed electrometer mainly comprises of a mechanically vibrating capacitor and an AC preamplifier, as shown in the conceptual schematic in Fig. 1. The time-varying capacitor modulates the weak static charge signal to the frequency higher than the noise corner frequency, where the flicker noise is much lower than the white noise. During the operation, a tiny amount of charge (or electrons) is induced to the immovable electrode of the parallel-plates of the vibrating capacitor, as shown in Fig. 1. The charge is then converted to an AC voltage V_i due to the movable electrode, which is

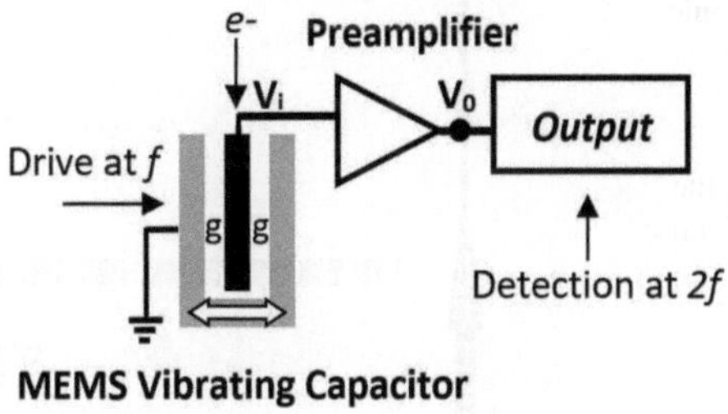

Fig. 1 Conceptual schematic diagram of the MEMS vibrating-reed electrometer. The main components include a mechanically vibrating capacitor, a low noise preamplifier and an output voltage detector at double frequency of driving signal (Jalil et al. ©IEEE [19])

vibrated by an external force of driving frequency f in a cyclic manner. Eventually, the AC signal from the output voltage of preamplifier V_0 is detected at $2f$ using signal measurement equipment.

For MEMS vibrating-reed electrometers, the moving plates of the vibrating capacitor is a mechanical resonator with a micro-meter scale traveling range yielding a relatively small electronic signal during measurement [25, 26]. In this sensing element in kHz range, the damping coefficient of the thermo-mechanical noise (or Brownian noise) yields a pico-meter range Brownian noise displacement, which corresponds to an infinitesimal electronic signal. Hence, Brownian noise has an insignificant amount of contribution to limit the charge resolution on the vibrating capacitor. On the contrary, the main obstacle in the realization of the high-resolution vibrating-reed charge sensors is their intrinsic electronic noise sources from the amplifying devices in the pre-amplifier, especially at low frequencies below the corner frequency of *1/f* noise, as shown in Fig. 2. At a higher frequency range above the corner frequency of *1/f* noise, both Johnson (thermal) and shot noises become dominant. Thus, minimizing the noise of the front-end electronics above the corner frequency of *1/f* noise is decisive in obtaining a high-resolution for charge sensors. Figure 2 illustrates the signal processing principle in frequency domain. The DC charge signal was initially hidden in large *1/f* noise and not possible to be detected at DC. Mechanical modulation shifts the DC signal to a higher frequency ($2f_m$), where the *1/f* noise was not dominant and the white noise is much less than the modulated charge signal, thereby improving the charge resolution.

Vibrating-reed electrometer utilizes a nonlinear variable capacitor driven by a balanced comb structure which up-modulates the input charge to twice the drive frequency prior to amplification, as shown in Fig. 3. Both the minimization of low frequency *1/f* noise and substantial rejection of charge feedthrough at the capacitor modulation frequency are the key advantages of the vibrating-reed principle [27, 28]. In the sense comb capacitor, the movable electrode plate is placed in the middle of two fixed electrode plates with same minimal gaps of g. Then the capacitance of the variable capacitor is given by

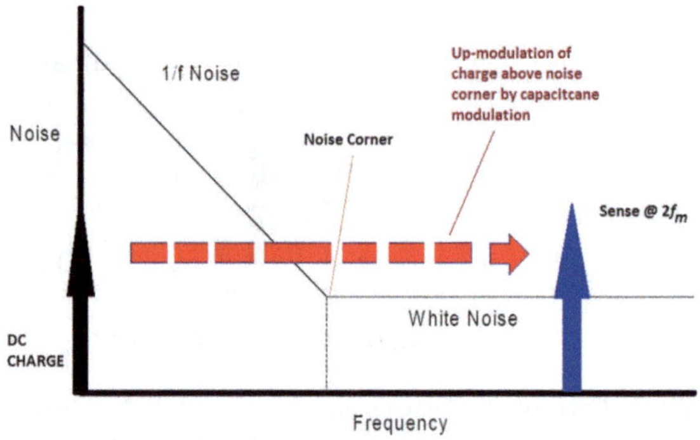

Fig. 2 Frequency characteristics of electronic noises in vibrating-reed electrometer. DC charge signal is modulated to a higher frequency above the corner frequency of *1/f* noise

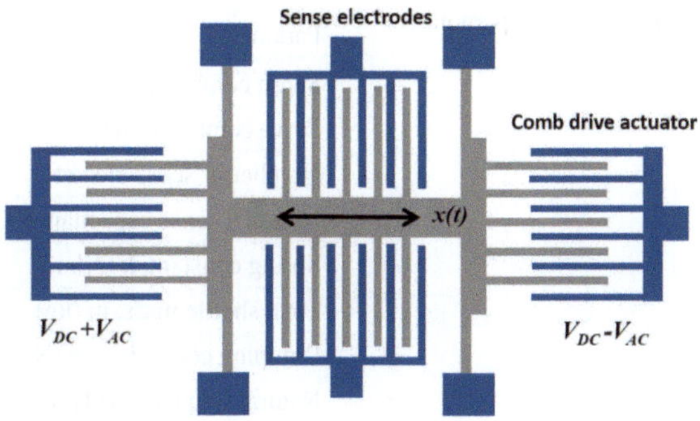

Fig. 3 Detailed schematic structures of the MEMS variable capacitor. Two comb drive actuators on both sides of the mechanical resonator actuate the shuttle in a push-pull manner

$$C_V(t) = C_0\left[\frac{g}{g + x(t)} + \frac{g}{g - x(t)}\right] \tag{1}$$

where C_0 is the nominal capacitance of the time-varying capacitor at the rest position, and

$$x(t) = \hat{x}\sin\omega t \tag{2}$$

where $\hat{x}$ is the maximum vibration displacement of the shuttle and ω is the vibration frequency of the shuttle. The vibration is excited by the push-pull comb actuators, and the frequency ω is the same as the actuation AC voltage. As shown in Eq. (1), due to the time-dependent gap, the capacitance C_v is also time-dependent. If higher harmonic (above 3rd) signals are ignored and a constant charge Q is applied on the time-varying capacitor C_v, the voltage on the capacitor is

$$V_i(t) = Q\left[\frac{1}{C_p + C_0} - \frac{\hat{x}^2}{g^2}\frac{C_0}{2(C_0 + C_p)^2} - \frac{\hat{x}^2}{g^2}\frac{C_0}{2(C_0 + C_p)^2}\cos 2\omega t\right] \tag{3}$$

where C_p is the total parasitic capacitance contributed by the connection pad and wires at the input terminal of the preamplifier. As this parasitic capacitance exists in all the denominators in Eq. (3), it should be kept as small as possible to have a large voltage signal. Equation (3) also indicates that the second harmonic signal at 2ω contains the DC charge information Q. Therefore, by detecting voltage signal at 2ω, one can reveal the weak DC charge information without the deterioration issue from flicker noise at the DC or low frequencies. In addition, detection at the twice modulation frequency can significantly suppress electrical drive feedthrough interference.

In addition, the charge sensitivity (or charge-to-voltage conversion gain) can be derived from Eq. (3):

$$\frac{dV}{dQ} = \frac{C_0}{2\sqrt{2}(C_0 + C_P)^2}\left(\frac{\hat{x}}{g}\right)^2 \tag{4}$$

Table 1 Design parameters and values

Parameter	Value
Sense comb gap, g (μm)	3
Drive comb gap (μm)	3
Number of sense electrode	42
Nominal sense capacitance, C_0 (pF)	0.7
Spring constant, k (N/m)	6.6
Total shuttle mass, m (kg)	4.8×10^{-9}
Damping coefficient, b (Ns/m)	3.1×10^{-5}
Natural frequency (kHz)	5.9
dC/dx of drive combs (F/m)	2.9×10^{-9}
DC drive voltage (V)	30
AC drive voltage (V)	10
Parasitic capacitance, C_p (pF)	36
Charge input capacitance, C_q (pF)	0.5
Structure thickness (μm)	10
Suspension beam width (μm)	3
Suspension beam length (μm)	300

As depicted in Eq. (4), the RMS magnitude of the second harmonic component of the voltage varies linearly with the input charge. The sensitivity can be increased by increasing the magnitude of vibration $\hat{x}$ and reducing the parasitic capacitance C_p, which provides a figure of merit guidance for the design optimization.

Table 1 shows the design parameters and values for the MEMS vibrating-reed electrometer. The values were determined by the above theoretical analysis and manufacturing constraints. According to the design values, the electrometer devices were fabricated with a commercially available multi-user SOI MEMS process through MEMSCAP. The microstructures were made by silicon and released by etching through the back of the substrate [29].

3 Simulink Modelling of the MEMS Electrometry System

3.1 Electrostatic Force Generation in Simulink

The vibrating-reed electrometer operates based on a mechanically resonating variable capacitor, which is excited by the electrostatic force F, as shown in Fig. 3. If the vibrating displacement is much smaller than the length of the beams, the following linear spring-mass-damper governing equation can be applied [30–32]:

$$m\ddot{x} + b\dot{x} + kx = F \tag{5}$$

where m is the total mass of the movable structure (including moveable electrodes of variable capacitor and the shuttle), b is the damping coefficient contributed mainly by the air viscous damping, and k is the mechanical spring constant of the clamped-clamped beams.

To generate push-pull driving forces to the shuttle, DC and AC voltages are superimposed and applied to the two comb actuators on both sides, as shown in Fig. 3. The values of driving voltages on each side of the shuttle are

$$V_1 = V_{dc} + V_{ac} \text{ and } V_2 = V_{dc} - V_{ac} \tag{6}$$

respectively. Therefore, the overall driving force applied to the shuttle is

$$F = \frac{1}{2}\frac{\partial C_d}{\partial x}V_1^2 - \frac{1}{2}\frac{\partial C_d}{\partial x}V_2^2 = 2\frac{\partial C_d}{\partial x}V_{dc}V_{ac} \tag{7}$$

where $\partial C_d/\partial x$ is determined by the geometrical parameters of the comb drive structures, including the gap, the thickness and the number of comb fingers. As depicted in Eq. (7), the actuation force can be tuned to the desired values by changing the comb structural dimensions during the design phase, and the DC and AC voltages during the testing phase.

To implement the actuation force numerically in Matlab Simulink, the constant block of V_{dc} is used to represent the value of DC voltage and the sine wave block V_{ac} represents the AC voltage in Eq. (7). The product block multiplies these two voltages together to get the value of $V_{dc}V_{ac}$, as shown in Fig. 4. The factor of $2\frac{\partial C_d}{\partial x}$ in Eq. (7) is implemented by cascading a gain block with a numerical gain of $2\frac{\partial C_d}{\partial x}$ as shown in Fig. 4. Then the output signal is fed into a scope to show the force waveform in time domain. As illustrated in the simulated result in Fig. 5, the generated electrostatic force has a sinusoid waveform with a magnitude of 1.72 μN and the frequency is found to be the same (5.9 kHz) as the AC driving voltage. This generated force has both periodically changed amplitude and direction, which can drive the shuttle to be vibrating and the details will be discussed in the following section.

Fig. 4 Simulink block diagram to simulate the electrostatic force generated by DC and AC actuation voltages

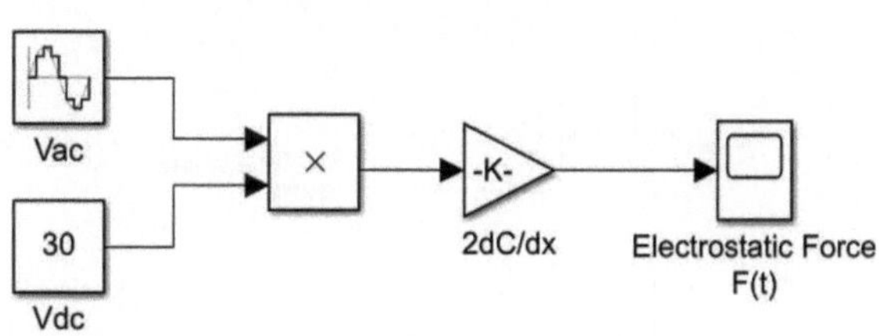

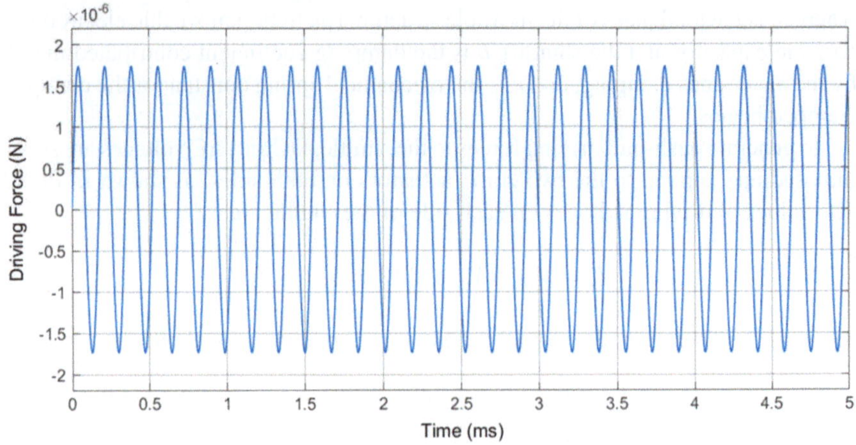

Fig. 5 Simulated waveform of the driving force on the shuttle, showing sinusoid wave force generated from the DC actuation voltage of 30 V and AC actuation voltage of 10 V at 5.9 kHz

3.2 *Equivalent Circuit Model of Mechanical Resonator*

With the obtained actuation force *F*, the shuttle vibrates periodically following the second order linear differential equation (Eq. 5). By contrast, a series inductor-capacitor-resistor (*LCR*) resonant circuit can also be described by a similar second order linear differential equation:

$$L\ddot{V}_x + R\dot{V}_x + \frac{V_x}{C} = \frac{V_{in}}{C} \tag{8}$$

where V_x is the output voltage across the capacitor *C*, and V_{in} is the input voltage applied to the *LCR* circuit as shown in Fig. 6. Comparing Eqs. (5) and (8), clear similarities can be found as they follow the same second order differential

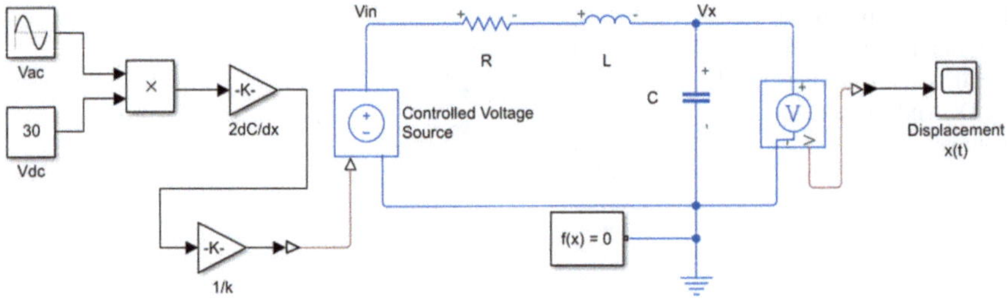

Fig. 6 Simulink implementation of the electrostatic actuation of the electrometer including the state equation of the electrostatic force generated by comb structures and equivalent *RLC* circuit representation of mechanical spring-mass-damper linear system

Table 2 Equivalences (substitutions) between variables in electrical and mechanical domains

Electrical quantities	Mechanical quantities
L (inductance)	m (mass)
R (resistance)	b (damping coefficient)
C (capacitance)	1/k (spring constant^{-1})
V_{in} (voltage source)	C × F (capacitance × driving force)
V_x (output voltage)	x (displacement)

governing equation. Therefore, numerical calculations can be done by proper substitutions of variables and unit conversions between the two equations. Table 1 details the equivalences between mechanical and electrical quantities. For example, as the time dependent variables in both equations, the output voltage V_x can represent the motional displacement *x*. After appropriate unit conversions, the variables can be used in the same numerical solver in Simulink. For instance, a shuttle displacement of 1 m is corresponding to an output voltage of 1 V; 1 kg mass is corresponding to an inductor with 1 H inductance, etc.

In addition, a voltage source of V_{in} in electrical domain is corresponding to a driving force *F* in mechanical domain with a scale factor of *C*, which is the capacitance in *RLC* circuit. As shown in Table 2, *C* is numerically equal to *1/k*, therefore,

$$V_{in} = CF = \frac{1}{k} \cdot F = \frac{1}{k} \cdot 2\frac{\partial C_d}{\partial x} V_{dc} V_{ac} \tag{9}$$

which links the actuation DC and AC voltages in mechanical domain to the input voltage of *RLC* circuit in electrical domain. By doing in this way, it is possible to put both mechanical and electrical components in the same simulator to do the system-level analysis. As shown in Fig. 6, a Gain block having a gain of 1/k is inserted between the Force output (shown in Fig. 4) and the *RLC* circuit to link mechanical and electrical domains. The output of the Gain block controls a *Controlled Voltage Source*, whose output value equals V_{in} in Eq. (9).

The output of the *Controlled Voltage Source* V_{in} excites the *RLC* circuit to resonant and the voltage V_x on the capacitor is corresponding to the shuttle displacement in value as shown in Table 2. The simulated displacement result is shown in Fig. 7, indicating that the displacement starts from 0 to a maximum of 1.25 μm within approximate 1.5 ms. After applying the driving force for 1.5 ms, the shuttle establishes a stable resonance with a displacement magnitude of 1.25 μm and a frequency of 5.9 kHz.

3.3 *Time-Domain Variable Capacitor Model in Simulink*

Having the time dependent displacement information in Fig. 7, it is possible to obtain the capacitance information over the time by using the capacitance Eq. (1). However, many commercially available circuit simulators, for example PSpice, do not accept in-line equations for capacitors. A complicated PSpice simulation technique has been

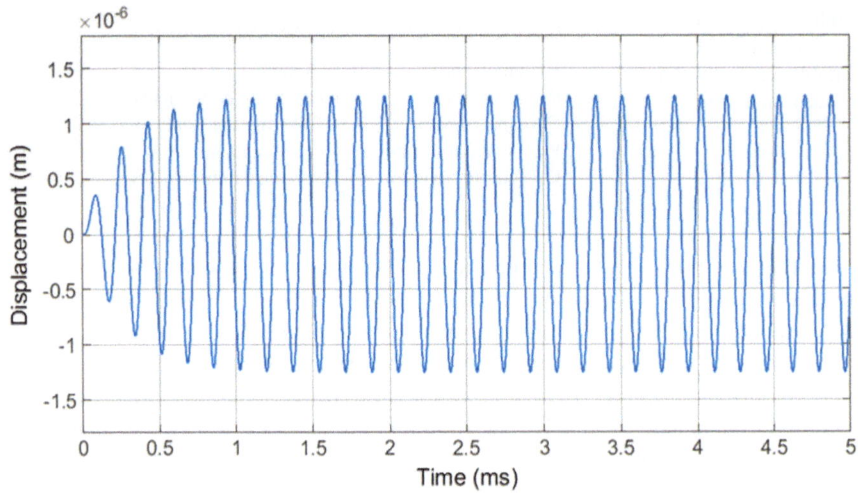

Fig. 7 Simulated waveform of the shuttle displacement, showing sinusoid wave displacement generated from the alternative electrostatic force

proposed to model a variable capacitor by a voltage-controlled voltage source, whose value is controlled by an auxiliary integrating circuit [6, 7]. Therefore, there is a need to directly model the time-varying capacitor in a more convenient circuit simulation environment. Alternately, Simulink in Matlab provides a much simpler way to model the time-dependent capacitor to be used in the circuit analysis solver in Simulink. It offers a circuit component block which can model a time-varying capacitor. The capacitance is set by the value of the physical input signal C, as shown in Fig. 8. The voltage and current relationship at the two terminals of the capacitor follows the following equation:

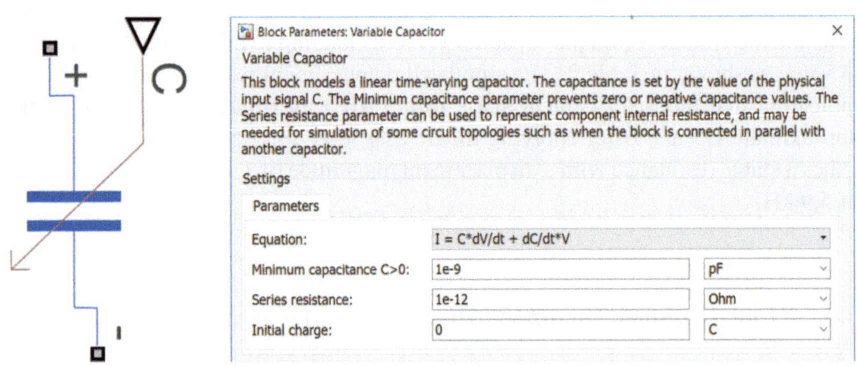

Fig. 8 Simulink variable capacitor block symbol and settings. The block has three terminals: '+' and '−' terminals connect to the electronic circuits, and 'C' terminal connects to the output of capacitance calculation blocks to set the time-varying capacitance values

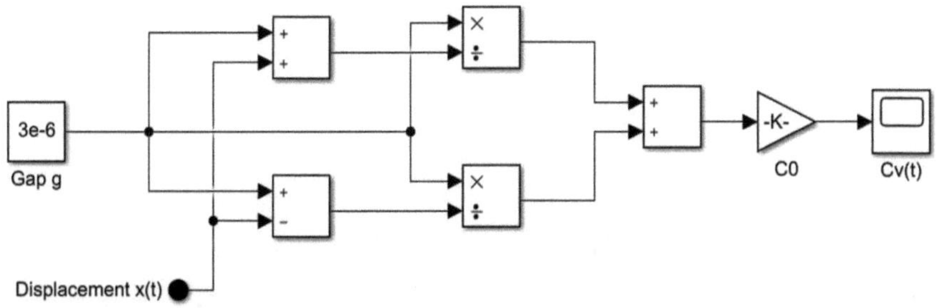

Fig. 9 Simulink block diagram implementation of the calculation of variable capacitance using Eq. (1). The time-varying displacement is fed from the displacement output in Fig. 6

$$I = C\dot{V} + V\dot{C} \tag{10}$$

Equation (10) ensures both voltage and capacitance's time-dependent information to be included in the calculation of the current through the capacitor. Therefore, the capacitance is also time-dependent, and the block can be used to simulate a circuit which has time-varying capacitors. In particular, the C terminal of the variable capacitor symbol connects to the output, $C_v(t)$ of the capacitance calculation blocks, as shown in Fig. 9.

To calculate the capacitance $C_v(t)$, Eq. (1) can be implemented in Simulink using generic mathematical calculation blocks of Add, Subtract, Divide, Gain and Constant. The connections are detailed in Fig. 9. To provide the displacement information to the capacitance calculation, the time-dependent displacement output in Fig. 6 is connected to the inputs of the Add and Subtract blocks.

Combining blocks in Figs. 6 and 9, capacitance can be simulated in Simulink, and the result is shown in Fig. 10. Due to the shuttle displacement, the gap between the fixed and moveable capacitor electrode plates are changing with time, thereby changing the capacitance. As seen in Fig. 10, the capacitance increases from the initial value of 1.4 pF to a maximum amplitude of 1.7 pF within approximate 1.5 ms. After 1.5 ms, the capacitance periodically varies with a frequency of 11.8 kHz, which is double of the frequency of the displacement.

3.4 Mechanical-Electrical System-Level Simulation

Equipped with the time-varying capacitance model in Simulink, the system-level simulation of the MEMS vibrating-reed electrometer can be achieved by integrating all blocks in Sects. 3.1, 3.2 and 3.3 with actual charge detection circuits. The complete Simulink implementation is illustrated in Fig. 11, which includes the force generation blocks (discussed in Sect. 3.1), the displacement blocks (discussed in Sect. 3.2), the capacitance calculation blocks (discussed in Sect. 3.3), and the actual charge detection circuits.

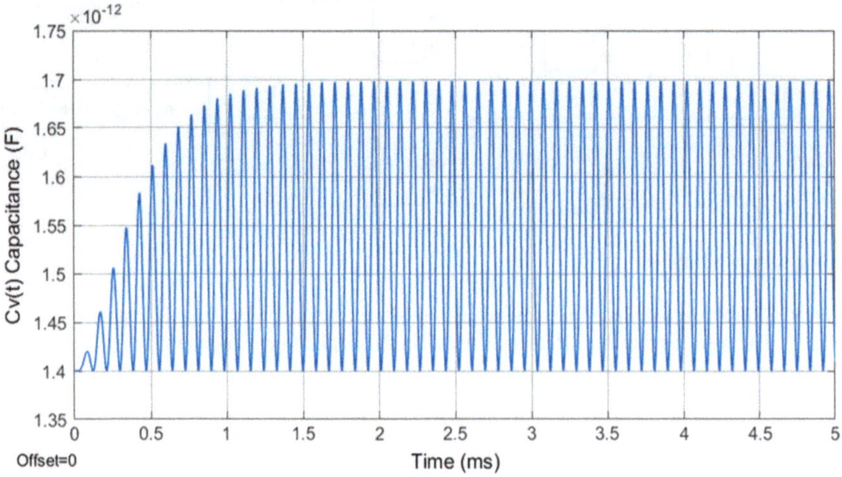

Fig. 10 Simulated waveform of the time-varying capacitance, showing a periodic wave over the time

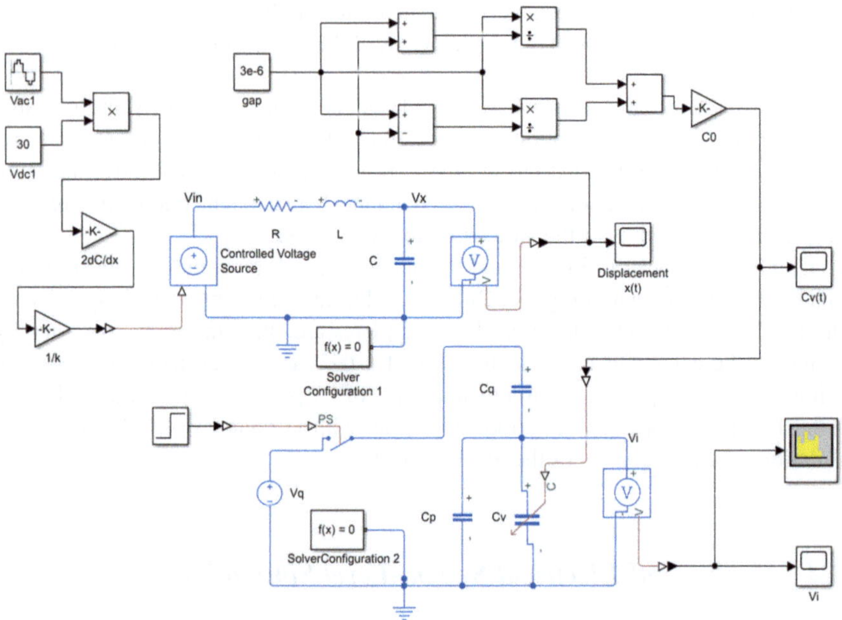

Fig. 11 A complete Simulink implementation of MEMS vibrating-reed electrometry system. Using a time dependent variable capacitor model in Simulink, the system-level simulation can be achieved to simulate both mechanical and electrical components in the same simulation environment

To apply known amount of charge to the time-varying capacitor C_v, a step voltage V_q is applied on a charge capacitor C_q, as illustrated in Fig. 11. The step transient voltage is achieved by connecting a switch to the DC voltage source of V_q. As the value of C_q is much less than the total value of C_v and parasitic capacitance C_p, the induced charge is approximately equal to $V_q \times C_q$. Having the charge induced on the variable capacitor, the electronic circuit transient simulation can be carried out in Simulink.

Figure 12 shows the waveforms of the voltage on the variable capacitor before and after introducing the charge. After charge is applied at 1 ms, the DC voltage is shifted to 0.33 V due to the voltage distribution among C_q, C_v and C_p. The small ripple voltage after 1 ms is contributed by the applied charge on the time-varying capacitor, which can be explained by second harmonic signal component in Eq. (3). To analyse the charge induced signal more accurately in frequency domain, the transient signal is fed in a Simulink block of Spectrum Analyzer, as shown in Fig. 11.

As illustrated in Fig. 13, a DC signal of −0.917 dBV exists in the spectrum, which is equal to the DC offset of 0.33 V in Fig. 12, and can be explained by the DC component in Eq. (3). The signal at 11.68 kHz has a magnitude of −60.95 dBV can be found in the spectrum, which is corresponding to the second harmonic component in Eq. (3). This signal is the detection signal in vibrating-reed electrometers as discussed in Sect. 2. The fourth harmonic is found at 23.35 kHz with a magnitude of −88.17 dBV, which is ~30 dB weaker than the second harmonic signal. It is why it has been ignored in Eq. (3), and is not used as the detection signal.

To characterize the charge-to-voltage conversion gain of the electrometer, the charge induced voltages are recorded when different input charges are introduced on the variable capacitor. As shown in the linear curve in Fig. 14, the conversion gain is 7.2×10^7 V/C, which matches well with the experimental results of 9.5×10^7 V/C

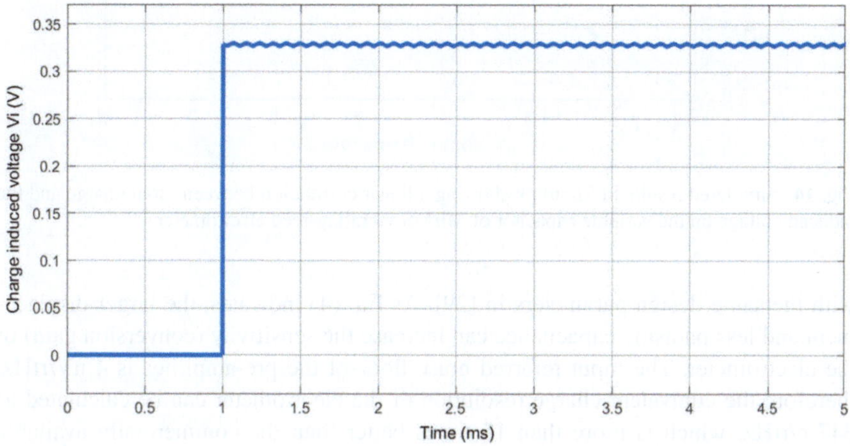

Fig. 12 The transient response waveform of the charge induced voltage. Known charge is introduced on the variable capacitor at 1 ms

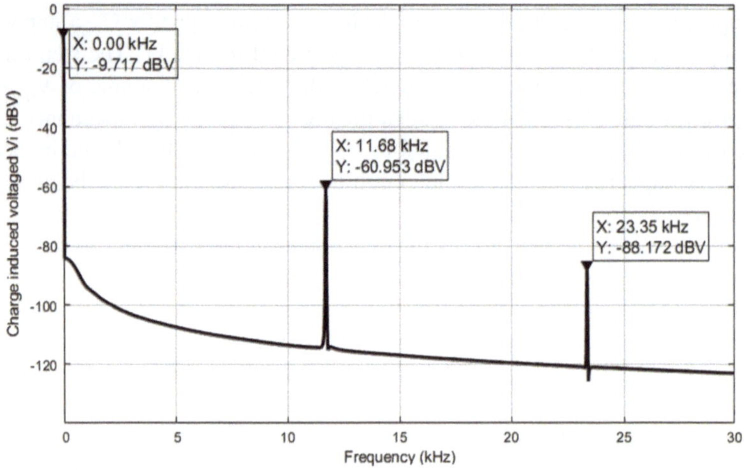

Fig. 13 Spectrum analysis of the charge induced voltage V_i. The DC, the second and the fourth harmonic signals can be seen in the frequency domain

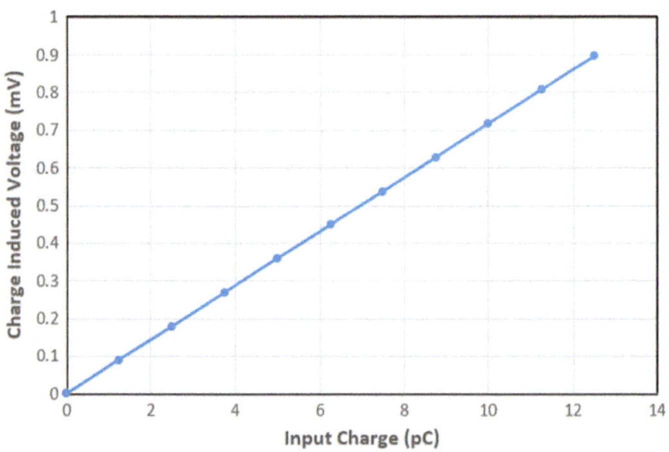

Fig. 14 Simulated results in Simulink showing a linear correlation between input charge and the induced voltage on the variable capacitor of MEMS vibrating-reed electrometer

with the same design parameters in [29]. As Eq. (4) indicates, the larger displacement and less parasitic capacitance can increase the sensitivity (conversion gain) of the electrometer. The input-referred noise floor of the pre-amplifier is 4 nV/rtHz, therefore the equivalent charge resolution of the electrometer can be calculated as 347 e/rtHz, which is more than 18 times better than the commercially available electrometer of Keysight B2987A [16].

4 Conclusions

This chapter proposed a novel and straightforward modelling approach to find the equivalent circuit representation of MEMS electrometer using time-dependent variable capacitor model in Matlab Simulink. The simulation results were in good agreement with the experimental results. This system-level simulation technique provides a simulation interface that combines both mechanical and electrical components and is a useful aid towards the modelling and optimization of the design of micro-electro-mechanical systems.

This simple Simulink-based technique of modelling time-dependent capacitor can be easily applied to any variable capacitor-based MEMS sensors, regardless of behavioural complexity of the capacitor. It can be achieved by simply revising the mathematical calculation blocks in Simulink for the capacitors with different time-dependence. Therefore, we hence presented a powerful tool that allows a fully combined simulation of both mechanical and electronic systems for the design and optimization of capacitive MEMS devices.

References

1. J.B. Angell, S.C. Terry, P.W. Barth, Silicon micromechanical devices. Sci. Am. **248**(4), 44–55 (1983)
2. Y. Zhu, A. Bazaei, S.O.R. Moheimani, M.R. Yuce, A micromachined nanopositioner with on-chip electrothermal actuation and sensing. IEEE Electron Device Lett. **31**(10), 1161–1163 (2010)
3. T. Veijola, H. Kuisma, J. Lahdenpera, T. Ryhanen, Equivalent circuit model of the squeezed gas film in a silicon accelerometer. Sens. Actuator A **48**, 239–248 (1995)
4. L.F. Che, B. Xiong, Y.L. Wang, System modeling of a vibratory micromachined gyroscope with bar structure. J. Micromech. Microeng. **13**, 65–71 (2003)
5. C. Basso, SPICE analog behavioral modeling of variable passives. Power Electron. Technol. 58 (2005)
6. Y. Zhu, J. Lee, A. Seshia, System-level simulation of a micromachined electrometer using a time-domain variable capacitor circuit model. J. Micromech. Microeng. **17**(5), 1059–1065 (2007)
7. Y. Zhu, J. Lee, A. Seshia, MEMS electrometer system simulation using a time-domain variable capacitor model, in *TRANSDUCERS 2007–2007 International Solid-State Sensors, Actuators and Microsystems Conference*, Lyon, 2007, pp. 1685–1688
8. H. Zhu, J. Lee, Simulating nonlinearity in MEMS resonators by a charge controlled capacitor. Procedia Eng. **25**, 403–406 (2011)
9. H. Zhu, J. Lee, System-level circuit simulation of nonlinearity in micromechanical resonators. Sens. Actuator A Phys. **186**, 15–20 (2012)
10. H. Paleosky, R.K. Swank, R. Grenchik, Design of dynamic condenser electrometers. Rev. Sci. Instrum. **18**(5), 298–314 (1947)
11. T.R. Ireland, N. Schram, P. Holden, P. Lanc, J. Vila, R. Armstrong, Y. Amelin, A. Latimore, D. Corrigan, S. Clement, J.J. Foster, W. Compston, Charge-mode electrometer measurements of S-isotopic compositions on SHRIMP-SI. Int. J. Mass Spectrom. **359**, 26–37 (2014)
12. J. Jalil, Y. Zhu, C. Ekanayake, Y. Ruan, Sensing of single electrons using micro and nano technologies: a review. Nanotechnology **28**(14), 142002 (2017)

13. G. Zimmerli, T.M. Giles, R.L. Kautz, J.M. Martinis, Noise in the Coulomb blockade electrometer. Appl. Phys. Lett. **61**, 237–239 (1992)
14. A.N. Cleland, M.L. Roukes, A nanometer-scale mechanical electrometer. Nature **392**, 160–162 (1998)
15. H. Kroemmer, A. Erbe, A. Tilke, S. Manus, R.H. Blick, Nanomechanical resonators operating as charge detectors in the nonlinear regime. Europhys. Lett. **50**, 101–106 (2000)
16. Keysight B2980A Series Femto/Picoammeter Electrometer/High Resistance Meter, Keysight Technologies [Online]. Available: https://literature.cdn.keysight.com/litweb/pdf/5991-4878EN.pdf?id=2500920. Accessed 28 Sept 2018
17. R.J. Schoelkopf, P. Wahlgren, A.A. Kozhevnikov, P. Delsing, D.E. Prober, The radio-frequency single-electron transistor (RF-SET): a fast and ultrasensitive electrometer. Science **280**(5367), 1238–1242 (1998)
18. I. Ahmed, J.A. Haigh, S. Schaal, S. Barraud, Y. Zhu, C.M. Lee, M. Amado, J.W.A. Robinson, A. Rossi, J.J.L. Morton, M.F. Gonzalez-Zalba, Radio-frequency capacitive gate-based sensing. Phys. Rev. Appl. **10**(1), 014018 (2018)
19. J. Jalil, Y. Ruan, H.Z. Li, Y. Zhu, Comprehensive design considerations and noise modeling of preamplifier for MEMS electrometry. IEEE Trans. Instrum. Meas. (2019). https://doi.org/10.1109/tim.2019.2930440
20. H. Palevsky, R.K. Swank, R. Grenchik, Design of dynamic condenser electrometers. Rev. Sci. Instrum. **18**, 298–314 (1947)
21. P.S. Riehl, K.L. Scott, R.S. Muller, R.T. Howe, J.A. Yasaitis, Electrostatic charge and field sensors based on micromechanical resonators. J. Microelectromech. Syst. **12**(5), 577–589 (2003)
22. J. Jalil, Y. Ruan, Y. Zhu, Room-temperature sensing of single electrons using vibrating-reed electrometer in silicon-on-glass technology. IEEE Electron Device Lett. **39**(12), 1928–1931 (2018)
23. G. Jaramillo, C. Buffa, M. Li, F.J. Brechtel, G. Langfelder, D.A. Horsley, MEMS electrometer with femtoampere resolution for aerosol particulate measurements. IEEE Sens. J. **13**(8), 2993–3000 (2013)
24. J.E. Lee, Y. Zhu, A.A. Seshia, A micromechanical electrometer approaching single-electron charge resolution at room temperature, in *2008 IEEE 21st International Conference on Micro Electro Mechanical Systems*, Wuhan, 2008, pp. 948–951
25. P.S. Riehl, K.L. Scott, R.S. Muller, R.T. Howe, High-resolution electrometer with micromechanical variable capacitor, in *Solid-State Sensor, Actuator and Microsystem Workshop (South Carolina)* (2002), pp. 305–308
26. J. Lee, Y. Zhu, A. Seshia, Room temperature electrometry with SUB-10 electron charge resolution. J. Micromech. Microeng. **18**(2), 025033 (2008)
27. Y. Zhu, J.E.Y. Lee, A.A. Seshia, A resonant micromachined electrostatic sensor. IEEE Sens. J. **8**(9), 1499–1505 (2008)
28. J. Jalil, Y. Zhu, T. Dinh, Y. Ruan, Development of a vibrating-reed MEMS charge sensor on silicon-on-glass technology, in *Proceedings of the 5th International Conference on Sustainable Design and Manufacturing (KES-SDM-18)*, Gold Coast, Australia, 2018, pp. 126–136
29. J. Lee, Y. Zhu, A. Seshia, Sub-10e charge resolution for room temperature electrometry, in *SENSORS, 2007 IEEE*, Atlanta, GA, 2007, pp. 821–824
30. J.E.Y. Lee, B. Bahreyni, A.A. Seshia, An axial strain modulated double-ended tuning fork electrometer. Sens. Actuator A Phys. **148**(2), 395–400 (2008)
31. D. Chen, J. Zhao, Y. Wang, Z. Xu, J. Xie, An electrostatic charge sensor based on micro resonator with sensing scheme of effective stiffness perturbation. J. Micromech. Microeng. **27**(6), 065002 (2017)
32. H. Zhang, W. Yuan, J. Huang, B. Li, H. Chang, A high-sensitivity micromechanical electrometer based on mode localization of two degree-of-freedom weakly coupled resonators. J. Microelectromech. Syst. **25**(5), 937–946 (2016)

MIX
Papier aus verantwortungsvollen Quellen
Paper from responsible sources
FSC® C105338

If you have any concerns about our products,
you can contact us on
ProductSafety@springernature.com

In case Publisher is established outside the EU,
the EU authorized representative is:
Springer Nature Customer Service Center GmbH
Europaplatz 3, 69115 Heidelberg, Germany

Printed by Libri Plureos GmbH
in Hamburg, Germany